Handbook of Conformal Mapping

with

Computer-Aided Visualization

V. I. Ivanov
M. K. Trubetskov
Physics Department, Moscow University

CRC Press
Boca Raton   Ann Arbor   London   Tokyo

Library of Congress Cataloging-in-Publication Data

Ivanov, V. IA. (Valentin IAkovlevich)
    Handbook of conformal mapping with computer-aided visualization / V. I. Ivanov,
M. K. Trubetskov.
        p.   cm.
    Includes bibliographical references and index.
    ISBN 0-8493-8936-4 (alk. paper)
    1. Conformal mapping—Data processing.   I. Trubetskov, M. K. (Michael K.)   II. Title.
QA360.I93   1994
515.9—dc20   94-34730
                                                                         CIP

Contents

To the memory
of the prominent mathematician,
our teacher
and the teacher of our teachers
Andrei N. Tikhonov

0.1 Historical Preface

A conformal mapping was first constructed and applied in cartography as the mapping of the surface of a sphere onto a plane with the smallest parts of the sphere being similar to their mappings. The simplest among conformal mappings of a sphere onto a plane — the stereographic projection — was described as long ago as the second century A.D. by Ptolemy. The founder of modern cartography, G. Mercator (1512–1594), published in 1568 the atlas of charts constructed in the conformal cylindrical projection that became the main one for sea navigation charts since then on.

The Lambert conic projection invented in 1772 by I. Lambert (1728–1777) is also a conformal although its author as well as his predecessors did not use this term.* It was first used in 1789 by Petersburg mathematician, astronomer and cartographer F. T. Schubert[1] (1758–1825).

Later, irrespective of Schubert, the term "conformal mapping" was used by C. F. Gauss (1777–1855) in his works devoted to cartography.[2] In this paper as well as in the early work devoted to geometry of surfaces[3] he investigated in detail the conformal mapping of one arbitrary surface onto another, in particular the mapping of the terrestrial spheroid onto a sphere and a plane. C. G. J. Jacobi (1804–1851) constructed the conformal mapping of the three-axis ellipsoid onto a sphere.[4] H. A. Schwarz (1843–1921) investigated the problems of conformal mapping of polyhedron surfaces onto a sphere (or a plane) and constructed a transformation of a tetrahedron onto a sphere.[5]

The modern theory of conformal mapping is a part of the theory of functions of a complex variable. The foundation of the theory of analytic functions was laid by L. Euler (1707–1783) who was the first[†] to obtain[6, 7] the conditions connecting the real and the imaginary parts of the analytic function $w = u + iv$ called Cauchy-Riemann (C-R) conditions:

$$\frac{\partial u}{\partial x} = \frac{\partial v}{\partial y}, \quad \frac{\partial u}{\partial y} = -\frac{\partial v}{\partial x}.$$

Euler has studied basic elementary functions of a complex variable in detail[9] and in particular has established the correspondence between exponential and trigonometric functions. He also considered power series with complex terms for the first time.[6]

After D'Alembert[8] L. Euler applied the theory of analytic functions to plane problems of hydrodynamics of ideal liquid.[6]

---

*However modern cartographers prefer the term "equiangular projection" instead of "conformal projection".

†Somewhat earlier C-R conditions were obtained by J. B. D'Alembert while studying the problem of plane vortex-free movement of liquid,[8] but he did not associate C-R conditions with the differentiability of a function of a complex variable.

L. Euler and J. L. Lagrange after him (1736–1813) first applied a complex variable to cartographic problems,[10, 11] namely to the transforms of conformal projections. They were the first to note the transition from one stereographic projection to another carried out with the help of linear-fractional function, parallels and meridians being mapped onto circumference or segments of a straight line.

O. L. Cauchy (1789–1857) founded the basics of strict theory of analytic functions in several works carried out in the first 20–30 years of XIX century (the theory of integrals of functions of a complex variable, the representation of analytic functions as contour integrals and power series, residues theory, etc.). A concept of analytic continuation playing the fundamental role in the theory of analytic functions has been introduced by C. Weierstrass (1815–1897).[12]

It was B. Riemann (1826–1866) who founded the geometric theory of analytic functions (i.e., the theory of a conformal mapping proper). His dissertation[13] gave birth to the broad use of ideas and methods of mathematical physics in the theory of functions. Riemann formulated the main theorem on existence of a conformal mapping of an arbitrary plane singly connected domain onto a circle. This theorem was named after him.[‡] Its proof, given by Riemann, was based on some variational principle of mathematical physics — the so-called Dirichlet principle. It assumes the existence of the minimum of some integral having the sense of field energy — the Dirichlet integral. This principle (and Riemann's proof as well) was criticized by C. Weierstrass (1815–1897) in the work[14] published in 1869 after Riemann's death. Weierstrass showed in an example that the minimum of Dirichlet integral was not possible to be reached on the set of differentiable functions. It was not until 1901, 50 years later than the work[13] published by Riemann when D. Hilbert (1862–1943) proved strictly the Dirichlet principle,[15] thus having rehabilitated Riemann's proof of this theorem.

Other proofs of the Riemann theorem (without the use of the Dirichlet principle) were given by H. A. Schwarz (1843–1921) and C. Caratheodory (1873–1950).[16, 17] The theorems of existence of conformal mappings of multi-connected domains were proved by P. Koebe (1882–1945) and D. Hilbert.[19] They proved that any N-connected domain could be conformally mapped onto a plane with N cuts along segments of parallel straight lines (the Hilbert's theorem), or onto a plane with N circles deleted (the Koebe's theorem).

J. Liouville (1809–1882) studied the conformal mapping of the domain of N-dimensional Euclidean space and has proved[18] that as $N > 2$ all

---

[‡]It is interesting to note that Riemann did not use the term "conformal mapping" yet, he spoke about "transform of the smallest parts of a plane into the similar ones". Concerning the mappings defined by the analytic functions the term "conformal mapping" was likely used for the first time by H. A. Schwarz.[16, 21, 22]

the conformal mappings could be reduced to linear transforms and to an inversion.[§]

The method of a construction of a conformal mapping of a half-plane (or a circle) onto an arbitrary polygon has been worked out independently by E. B. Christoffel (1829–1900) and H. A. Schwarz.[20, 21] H. A. Schwarz also constructed the conformal mapping of a circle onto an ellipse (with the help of elliptic functions)[22] as well as on a circular polygon (with the help of the so-called automorphic functions).[21, 23] The theory of elliptic functions was developed by many scientists, first of all by C. G. J. Jacobi and C. Weierstrass.[24–27] The theory of automorphic functions has been created by works of F. Klein (1849–1925) and H. Poincare (1854–1912).[28–31]

A remarkable theorem was formulated by P. L. Chebyshev (1821–1894):[32] "Among all the conformal projections of the given domain belonging to a sphere (a country, a continent or an ocean) onto a plane (the map) the best one is that with the constant scale of the image in the points of the boundary of the domain". The proof of this theorem was given in 1894 by Russian mathematician L. A. Grave (1863–1939).[33]

After B. Riemann has associated the problems of conformal mappings with the boundary-value problems of mathematical physics[13, 34] intensive work began on developing applications of conformal mappings to various fields of physics.

H. Helmholtz (1821–1894) and G. R. Kirchhoff (1824–1887) used conformal mappings to solve plane problems of hydrodynamics of ideal liquid, in particular the problems of a jet stream of liquid.[35, 36]

G. R. Kirchhoff was the first to apply conformal mappings to plane problems of electrostatics and magnetostatics as well as to the problems of the distribution of electric current on a plane plate and on a curved surface.[37, 38]

N. E. Zhukovskii (1847–1921) and S. A. Chaplygin (1869–1942) applied conformal mappings to problems of hydro- and aeromechanics, in particular to the theory of a wing,[39–41] to the theory of a jet,[42] and to the theory of flow around grids.[42–44] Conformal mappings of periodic structures (grids) were also studied by N. E. Kochin (1901–1944).[45]

G. V. Kolosov (1867–1934) and N. I. Muskhelishvili (1891–1976) gave birth to the broad use of conformal mappings in the theory of elasticity.[46–48]

N. N. Pavlovskii (1884–1937) was the first to use conformal mappings in problems of filtration, that is, the movement of ground water under hydrotechnical constructions.[49]

M. A. Lavrent'ev (1900–1980), M. V. Keldysh (1911–1978), and L. I. Sedov (1907–) applied conformal mappings to the problem of the impact of

---

[§]Therefore the class of conformal mappings of N-dimensional space with $N > 2$ is rather poor. In this Handbook such mappings are not considered.

a solid cylinder on the surface of a noncompressible liquid.[50, 51]

Numerical methods of conformal mappings were developed by M. A. Lavrent'ev, L. V. Kantorovitch (1912–1986), V. I. Krylov (1902–), P. F. Fil'chakov (1916–1978) and many others.[52–55]

0.2 Introduction

A conformal mapping is a transform of geometric figures in which infinitely small pieces are transformed into similar ones. The theory of conformal mappings of plane domains is closely associated with the theory of analytic functions of a complex variable. An analytic function considered as a mapping defines (under certain conditions) a conformal mapping of a domain of its specification onto a domain of its values. Any person studying the theory of functions of a complex variable comes across conformal mappings as graphic illustrations of analytic functions. The theory of a conformal mapping can be considered an essential part of mathematical education.

On the other hand, the importance of conformal mappings arises from their numerous applications in such fields as hydro- and aerodynamics, the theory of filtration, the theory of elasticity, the theory of magnetic and electric fields, heat conductivity and cartography. A great number of specialists in these fields have to use conformal mappings when calculating physical fields.

A wide and rich literature is available on conformal mappings including reference guides (see the bibliography at the end of the book). Theoretical aspects of conformal mappings are considered in all the textbooks on the theory of analytic functions for mathematicians.[56-62] General mathematical questions on conformal mappings are considered in the books.[63-72] Applications of conformal mappings to physical problems are studied in the textbooks[73-87] written for students in physics and engineers. General applications of conformal mappings are discussed in References.[101-108] The application of conformal mappings to the problems of hydrodynamics are discussed in References[109-131], in particular, References[114-119] are devoted to problems on streaming around aerofoils, References[120-126] — to problems on streaming around grids, References[127-129] — to problems on filtration of ground water under hydrotechnical constructions, References[130, 131] — to problems on jet streams. Conformal mappings applied to calculation of electrical and magnetic fields, to the elasticity and to heat transfer are discussed in References[132-146], References[147-162] and Reference[163], correspondingly. A number of books of problems[164-169] and handbooks[170-177] including problems on conformal mappings are available. All the books were written in the pre-computer epoch.

Progress in computing and graphical performance of personal computers has significantly expanded the possibility of conformal mapping construction and application. But till now there has not been created any manual on the visualization of a conformal mapping on a display. The aim of this book is to fill the gap.

The work is a reference-book on conformal mappings, and their application and construction on a display. It is designed for persons studying the theory of analytic functions and their applications, as well as for investigators and engineers using conformal mappings in hydro- and aerodynamics, and the theory of electric and magnetic fields.

Since this book is oriented to students, the bibliography does not include journal articles but only books (except for the literature on the Introduction). As it is also for specialists in applied mathematics, theorems on the properties of a conformal mapping are given without proofs.

The book consists of three Chapters:

1. Mathematics (the Theory of Conformal Mappings)

2. Physics (Applications of Conformal Mappings)

3. Practice (the Catalogue of Conformal Mappings)

The diskette enclosed contains the interactive program CONFORM allowing one to build all the mappings performed in the book and to construct the new ones. This program should be run under Microsoft Windows* version 3.1 or higher on IBM-compatible PC. It includes the Database system, Interpreter of functions of a complex variable and Visualizer of mappings. The installation and user's guides are given in the Appendix.

We hope that the book and program CONFORM will make a study of conformal mappings not only effective but also pleasant.

---

*Windows is a trademark of the Microsoft Corporation.

0.3 Acknowledgments

The authors are sincerely thankful to the professor of Moscow State University Aleksei G. Sveshnikov, who has read the manuscript and has made a lot of helpful remarks. The authors are also grateful to many colleagues from Physics Department of MSU for discussions and support, in particular — to Aleksei D. Poezd and Fedor V. Shugaev. The authors are thankful to University of Dallas professor Richard P. Olenick for the benevolent reference.

The authors also want to thank Lidia A. Korchunova for the help in the printing of the manuscript and to Marina D. Trubetskova for the help in its translation.

1

Mathematics: the Theory of Conformal Mappings

1.1 Complex Plane, Domains and Curves on It

We assume the reader of "Handbook of Conformal Mapping with Computer-Aided Visualization" is acquainted with the basis of the theory of functions of a complex variable, but nevertheless we find it necessary to review the basic concepts briefly.

1.1.1 The complex plane and the sphere of complex numbers

A *complex number* is an ordered pair of two real numbers (a, b) that we shall designate as $z = a + ib$, where i is the imaginary unit, a is called the real part of the complex number z and b is its imaginary part: $a = \operatorname{Re} z$, $b = \operatorname{Im} z$. Complex numbers can be added and multiplied in accordance with the rules of addition and multiplication of polynomials, taking into account that $i^2 = -1$.

A complex number $z = a + ib$ can be depicted by a point in a plane with Cartesian coordinates a, b. This plane (and correspondingly the set of all complex numbers) is designated as C. The axis of abscissae of the complex plane is called the real axis, the axis of ordinates — the imaginary one.

The position of the point z in a plane can be characterized (besides the Cartesian coordinates) by polar coordinates ρ, φ as well:

$$a = \rho \cos \varphi, \quad b = \rho \sin \varphi;$$

$$z = \rho \left(\cos \varphi + i \sin \varphi \right). \tag{1.1.1}$$

The value ρ is called the *modulus of a complex number* z; it represents the distance between the point $z = a + ib$ and the origin of the plane:

$$\rho = |z| = \sqrt{a^2 + b^2}.$$

The value $|z - z_0|$ represents the distance between the points z_0 and z. The value φ is called the *argument of a complex number z* (Arg z), it represents the angle between the radius-vector of the point z and the positive direction of the x-axis. φ is defined by the system of equations

$$\cos \varphi = x/\rho,$$

$$\sin \varphi = y/\rho.$$

The angle φ is not defined uniquely by this system: its value can vary by $2\pi n$, where $n \in \mathcal{Z}$, $\mathcal{Z}$ is a set of integer numbers. Within the interval $-\pi < \varphi \leq \pi$ the system has exactly one root called the *principal value of the argument* and designated as $\arg z$, so

$$\varphi = \text{Arg } z = \arg z + 2\pi n.$$

The argument of the point $z = 0$ is not defined. For the rest of the points in the plane we have

$$\arg z = \begin{cases} \arctan(b/a), & a > 0, \\ \arctan(b/a) + \pi, & a < 0, \quad b \geq 0, \\ \arctan(b/a) - \pi, & a < 0, \quad b < 0, \\ \pi/2, & a = 0, \quad b > 0, \\ -\pi/2, & a = 0, \quad b < 0. \end{cases}$$

The number $\bar{z} = a - ib$ is called the *complex conjugate* for the complex number $z = a + ib$. It is depicted by the point symmetrical to the point z with respect to the real axis (Fig. 1.1) with

$$|\bar{z}| = |z|, \quad \arg \bar{z} = -\arg z.$$

A complex number z designated as $z = a + ib$ is called the algebraic form and that designated as (1.1.1) is called the trigonometric form of a complex number. With the help of *Euler formula*

$$e^{i\varphi} = \cos \varphi + i \sin \varphi \tag{1.1.2}$$

the expression (1.1.1) can be written somewhat shorter:

$$z = \rho e^{i\varphi}.$$

It is called the exponential form of a complex number.

Complex numbers are more conveniently added and subtracted in algebraic form. Multiplication, division and raising to an integer power is more convenient if complex numbers are represented in trigonometric (or exponential) form. If

$$z_1 = \rho_1 e^{i\varphi_1}, \quad z_2 = \rho_2 e^{i\varphi_2},$$

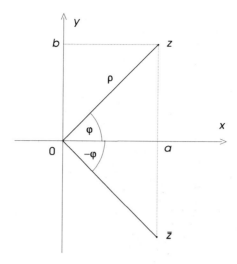

FIGURE 1.1

then

$$z_1 z_2 = \rho_1 \rho_2 e^{i(\varphi_1 + \varphi_2)}, \quad \frac{z_1}{z_2} = \frac{\rho_1}{\rho_2} e^{i(\varphi_1 - \varphi_2)}, \quad z_1^n = \rho_1^n e^{in\varphi_1}.$$

There exists another geometrical interpretation of the set of all the complex numbers (besides the complex plane $\mathcal{C}$): the *complex numerical sphere* or Riemann sphere. Reciprocal one-to-one correspondence of the points in the plane and that of the sphere is determined with the help of *stereographic projection*.

To construct the stereographic projection let us consider the plane $\mathcal{C}$ as a coordinate plane $\zeta = 0$ in three-dimensional space $\mathcal{R}^3$ with Cartesian coordinates ξ, η, ζ. Let us construct the sphere S with the unit radius whose center is the point $(0, 0, -1)$ touching the plane in its origin. Let us call as the pole of the sphere the end of its diameter passing through the point of the touch. Designate the pole by P; its Cartesian coordinates are $(0, 0, -2)$. The stereographic projection of the point z in the plane onto the S-sphere is the point M of the crossing of the straight line connecting the point z and the pole P with the sphere S (Fig. 1.2). Obviously, each point on the sphere not coinciding with the pole P is mapped reciprocally one-to-one onto some point of the $\mathcal{C}$ plane. To make the mapping of the entire sphere S onto the plane reciprocal one-to-one, let us introduce the concept of the *point at infinity* as the image of the pole P in the case of stereographic projection of the sphere onto a plane.

A complex plane with the point at infinity added is called the *extended complex plane* and is designated as $\overline{\mathcal{C}}$ (sometimes it is called the complex sphere). We designate the point at infinity of a plane as ∞. The $\mathcal{C}$-plane it-

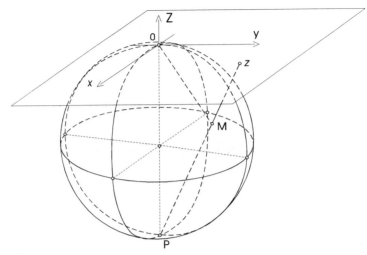

FIGURE 1.2

self is sometimes called the open complex plane, the point $z = \infty$ represents its boundary.

Let us introduce spherical coordinates on the sphere S:

$$\xi = \sin\theta\cos\varphi,$$

$$\eta = \sin\theta\sin\varphi,$$

$$\zeta = -1 + \cos\theta,$$

where $0 < \theta < \pi$, $-\pi < \varphi \leq \pi$. The stereographic projection of the point in the sphere S with spherical coordinates θ, φ is the point $z = 2\tan(\theta/2)e^{i\varphi}$. Inversely, the stereographic projection of the point in the plane $z = \rho e^{i\varphi}$ is the point in the S-sphere with spherical coordinates θ, φ, where

$$\theta = 2\arctan(\rho/2).$$

1.1.2   Domains in the complex plane

The set of points in the plane is called *connected* if any two of its points can be connected by a curve entirely belonging to this set.

We call a *neighborhood* (ε-neighborhood) of the point z_0 ($z_0 \in C$) the set of points in the complex z-plane, satisfying the condition $|z - z_0| < \varepsilon$, where $\varepsilon > 0$. The set of points satisfying the condition $0 < |z - z_0| < \varepsilon$ is called a punctured neighborhood of the point z_0.

We call the neighborhood of the *point at infinity* $z = \infty$ (or R-neighborhood of the point at infinity) the set of points z, satisfying the condition $|z| > R$, where $R > 0$.

Let D be some set of points in a complex plane. The point z is called the *interior point* of the set D if there exists its (rather small) ε-neighborhood, such that all its points belong to D.

The set D is called a *domain* if:

1. this set is a connected one,

2. all its points are interior.

A finite point z is called the *boundary point of the domain D* if it does not belong to the domain but in any (as small as desired) ε-neighborhood of it there exist points that belong to D.

The finite point z is called the *isolated boundary point of the domain D* if it does not belong to D but some of its (small enough) punctured ε-neighborhood belongs to the domain D.

The finite point z is called the *exterior point of the domain D* if there exists ε-neighborhood such that none of its points belong to the domain D.

The given definitions of the interior, exterior and boundary finite points can be extended on the case of the point at infinity substituting for ε-neighborhood of the finite point the R-neighborhood of the point at infinity.

The set of all boundary points of the domain D is called the *boundary of the domain*; we denote it as ∂D. The extended complex plane has no boundary; other domains of the complex plane have a boundary containing at least one point.

The domain D joined with its boundary ∂D is called a *closed domain*; we designate it by $\overline{D}$, therefore

$$\overline{D} = D \cup \partial D.$$

Example 1.1

The set of all of the points of the square $0 < x < 1, 0 < y < 1$ represents the domain. The set of all of the points of the square with rational coordinates $x = m/n, y = k/l, (m, n, k, l \in \mathcal{N}, m < n, k < l)$ does not represent a domain because it is not connected and its points are not the interior ones. Here $\mathcal{N}$ designates a set of natural numbers.

Example 1.2

The set of points z, satisfying the condition $|z - z_0| < R$ (where $R > 0$), represents the domain — the circle with the center at the point z_0 and with the radius R. The circumference $|z - z_0| = R$ is its boundary.

Example 1.3

The set of points z satisfying the condition $|z - z_0| \leq R$ $(R > 0)$ is a closed domain.

Example 1.4

The set of points z satisfying the condition $|z - z_0| \neq R$ $(R > 0)$ is not a domain, because this set is not connected.

1.1.3   Curves in a complex plane

Let L be a curve in the $\mathcal{C}$-plane defined by the parametric equation:

$$x = x(t), \quad y = y(t), \tag{1.1.3}$$

where $x(t)$, $y(t)$ are real-valued continuous functions of the real parameter t, given on the segment $a \leq t \leq b$. We shall also denote the curve L in the different form:

$$z = z(t), \quad a \leq t \leq b, \tag{1.1.4}$$

where $z(t) = x(t) + iy(t)$ is the complex-valued function of the real parameter t. Let us assume that the function $z(t)$ is such that $z(t_1) \neq z(t_2)$ as $t_1 \neq t_2$ (where t_1, t_2 are any points on the segment $[a, b]$), i.e., the curve L does not have any self-intersection point. In this case the curve (1.1.4) is called a *Jordan curve*.

If the beginning of a Jordan curve coincides with its end ($z(a) = z(b)$), then such a curve is called the *closed Jordan curve*, the coincidence of the initial and the terminal points of the curve is not considered as a self-intersection.

The Jordan curve (1.1.4) is the image of the segment $[a, b]$ in the case of the continuous single-valued mapping of it onto a plane (such a mapping is called the homeomorphic one). The closed Jordan curve can be considered as the image of a circumference in the case of homeomorphic mapping of a circumference onto the $\mathcal{C}$-plane.*

The closed Jordan curve on the complex sphere (on the extended complex plane $\overline{\mathcal{C}}$) passing through the point at infinity is called the *infinite Jordan curve*. It can be considered as the image of an infinite straight line in the homeomorphic mapping onto the plane $\overline{\mathcal{C}}$.

The curve (1.1.3–1.1.4) is called *smooth* if there exists a tangent to any of its points. From a course in mathematical analysis it is known that, for

---

*Indeed for the closed Jordan curve a parameter $\tau = 2\pi t/(b - a)$ can be introduced and interpreted as a polar angle on the circumference.

existence of a tangent, it is sufficient that the derivatives $\dot{x}(t)$ and $\dot{y}(t)$ exist
to satisfy the condition $\dot{x}^2(t)+\dot{y}^2(t) \neq 0$ or else $\dot{z}(t) \neq 0$. An infinite Jordan
curve is called smooth at the point at infinity if there exists $\lim \arg \dot{z}(t)$
as $|t| \to \infty$. We shall consider only domains with boundaries representing
piecewise smooth curves, i.e., curves consisting of a finite number of smooth
curves.

1.1.4   The classification of plane domains

The domain D in the plane $\mathcal{C}$ is said to be *finite* if there exists a circle such
that it contains the domain D. Otherwise the domain is called an infinite
one.

Let L be a closed Jordan curve in the extended complex plane $\overline{\mathcal{C}}$. This
curve divides the plane $\overline{\mathcal{C}}$ into two domains: the interior domain D_i and
the exterior one D_e, the curve L being their boundary.[†]

The finite domain D_i has a remarkable property: any closed Jordan
curve belonging to D_i bounds the domain that also entirely belongs to
D_i. In other words, any contour lying in D_i can be tightened into a point
belonging to D_i by continuous deformation without going outside D_i. The
infinite domain D_e possesses a similar property: any closed curve lying in
D_e divides the extended plane $\overline{\mathcal{C}}$ into two parts, either an interior part or
an exterior one belonging to D_e. In the domain D_e as well any contour
can be tightened by continuous deformation, not exceeding the bounds of
D_e into a finite point or a point at infinity.

The domains having such a property are called *singly connected* ones.
The domain D_i is an example of the finite singly connected domain; for it
the point $z = \infty$ is exterior. The domain D_e is an example of an infinite
singly connected domain; we shall name it as an exterior of the contour L.
For D_e the point at infinity is interior.[‡]

The point $z = \infty$ can be the boundary point of the singly connected
domain as well. For example, let us consider the infinite curve L without
any self-intersection, defined by parametric equation:

$$z = z(t), \quad t \in \mathcal{R}, \tag{1.1.5}$$

where $z(t)$ is a continuous function of a real parameter t unlimitedly increasing as $t \to \pm\infty$. We call the curve (1.1.5) the closed Jordan curve
on the complex sphere. Let this curve as $t \to +\infty$ and $t \to -\infty$ have
asymptotes with the angle α between them (Fig. 1.3). The curve (1.1.5)
divides the complex plane $\mathcal{C}$ into two infinite singly connected domains D_1

[†]This statement is called Jordan's theorem. At the first glance it seems evident but
to prove it strictly is not easy.

[‡]Note that by excluding the point $z = \infty$ the domain D_e would be doubly connected.

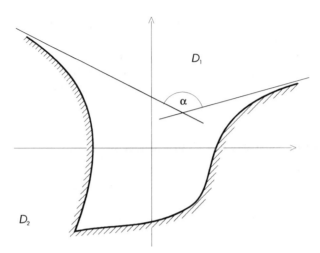

FIGURE 1.3

and D_2, we name them *curvilinear angular domains* with apex angles α
and $2\pi - \alpha$ correspondingly. In the case of $\alpha = \pi$ we name the domains
D_1 and D_2 as curvilinear half-planes. For curvilinear angular domains D_1
and D_2 the point $z = \infty$ is the boundary point.

One of the most important characteristics of a plane domain is the order
of connectedness.

In the case of a finite domain with the boundary consisting of the finite
number of closed curves, cuts and isolated singular points the *order of
connectedness* is the number N of connected parts of the boundary.[§] The
domain itself is called the N-connected one. For example, in Fig. 1.4 there
is depicted a finite domain with the boundary consisting of the two closed
curves L_0 and L_1, the two cuts γ_1 and γ_2 and the isolated boundary point
z_1. This domain is the multiply connected one ($N = 4$).

In the case of an infinite domain the order of connectedness N is defined
as the number of connected sections of the boundary of the domain. If the
point $z = \infty$ is an isolated boundary point of the domain it is accounted
for as a connected set when calculating the order of the connectedness. If
the point $z = \infty$ is an unisolated boundary point and the boundary of the
domain includes several infinite curves passing through the point $z = \infty$,
all of them are accounted for as one connected part of the boundary.

To classify plane domains the concept of multiplicity of the boundary
points of a domain is useful. It is convenient to introduce this concept first
for the finite singly connected domain (Fig. 1.4). Let us choose some initial
point and traverse the boundary starting from this point in the positive

---

[§]In this case each isolated boundary point is considered to be a connected set.

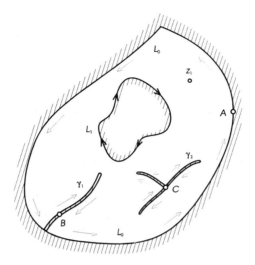

FIGURE 1.4

direction (the positive direction of the boundary traversal is that when the domain always remains to the left). In case of the full boundary traversal some of its points are traversed once, some of them twice, three times, etc. Correspondingly the boundary points are called one-fold, two-fold, three-fold ones, etc. For example, in Fig. 1.4 the point A is a one-fold point, B — a two-fold one, and C — a three-fold point.

This definition is also suitable for multiply connected domains as well as for the infinite ones. In particular, the point at infinity is the K-fold boundary point of some domain if, in the case of the full boundary traversal of the domain, the observer passes through the point at infinity K times. The boundary of such a domain consists of K infinite non-intersecting curves.

The concept of the curvilinear angular domain was introduced above. The point $z = \infty$ is its one-fold boundary point. The other kind of the infinite singly connected domain with the infinite boundary is the domain concluded between two non-intersecting infinite curves. We call such a domain a *curvilinear strip*. For such domains the point $z = \infty$ is the two-fold boundary point.

In the general case the boundary of the singly connected infinite domain can consist of K infinite curves. We call such domains *curvilinear strips with K branches*. The example of such a domain is the plane C with K cuts along the non-intersecting straight-line rays.

Therefore, plane domains can be classified in such a way: singly connected and multiply connected domains, finite and infinite ones, and domains with finite and infinite boundaries. The domains with infinite boundaries can be classified according to the multiplicity of the point at infinity.

The complete classification of the singly connected domains is given below:

1. The extended complex plane $\overline{C}$ — the domain without any boundary.

2. The extended complex plane $\overline{C}$ with one point deleted (finite or infinite one) — the domain with the boundary consisting of a single point.

3. The finite domain. As a rule its boundary is a piecewise smooth closed curve, we name it as a *contour*. If all the points of the contour are one-fold boundary points, then this contour is the Jordan curve. The point at infinity is the exterior point for the finite domain.

4. The infinite domain with the finite boundary. The boundary of the domain is the closed piecewise smooth curve, we also name the curve as the contour, and the domain as the *exterior of the contour*. If all the points of the contour are one-fold boundary points, then it represents the closed Jordan curve. The point at infinity is the interior one for this domain.

5. The curvilinear angular domain (in particular curvilinear half-plane). The point at infinity is one-fold boundary point for such a domain.

6. The curvilinear strip. The point at infinity is a two-fold boundary point for such a domain.

7. The curvilinear strip with N branches ($N > 2$). The point $z = \infty$ is the N-order boundary point for such a domain.

Singly connected domains of the 1 and 2 type are called the *degenerate* ones, and the domains of types 3–7 the nondegenerate ones. The boundary of the nondegenerate domains contains more than one point.

In the following chapters of the Handbook we consider the conformal mapping of different domains onto some domains of the simple form called canonical domains. Each of the singly connected domains of type 1–7 has the corresponding canonical domain, degenerate domains being able to be mapped only onto the degenerate ones.

1. The extended complex plane $\overline{C}$ will be shown to be able to be mapped conformally only onto the extended complex plane $\overline{C}$. The extended complex plane $\overline{C}$ is represented by the canonical domain itself.

2. For the extended complex plane $\overline{C}$ with one point deleted the canonical domain is represented by the open complex plane C.

3. For finite singly connected domains the canonical domain is represented by the unit circle $|z| < 1$.

4. For infinite singly connected domains with a finite boundary (i.e., for the exteriors of finite contours) the canonical domain is represented by the exterior of the unit circle $|z| > 1$ with the point at infinity included.

5. For the curvilinear angular domains (in particular — for the curvilinear half-planes) the canonical domain is represented by a half-plane Im $z > 0$.

6. For the curvilinear strip the straight line horizontal strip $0 < $ Im $z < \pi$ is regarded as a canonical domain.

7. For the curvilinear strip with N branches the plane $\mathcal{C}$ mentioned above with cuts along N non-intersecting straight line rays can be taken as a canonical domain.

The complete classification of doubly connected domains can be made considering that doubly connected domains result from the singly connected ones of types 2–7 by deleting either one point or any closed subdomain or by making a cut along a finite open curve (for example, Jordan curve), entirely belonging to the domain. We shall confine ourselves to the most general classification of doubly connected domains according to the degeneration of boundaries. There are three types of them:

1. The extended complex plane $\overline{\mathcal{C}}$ with two points z_1 and z_2 deleted. In this case both of the two connected parts of the boundary are degenerate into points. This domain can be conformally mapped only onto the extended complex plane $\overline{\mathcal{C}}$ with two points deleted. The canonical domain in this case can be represented by the open complex plane $\mathcal{C}$ with point $z = 0$ deleted.

2. A nondegenerate singly connected domain with an interior point deleted represents a doubly connected domain having one of the two connected parts of the boundary degenerate. A unit circle $0 < |z| < 1$ with the center deleted is usually concerned as the canonical doubly connected domain of such a type.

3. All the other doubly connected domains (with nondegenerate boundaries) can be obtained by excluding from the nondegenerate singly connected domain of either the point or the cut drawn through the open curve or some closed domain. An annulus $1 < |z| < R$ is usually taken as a canonical domain for doubly connected domains of such a type, where $R > 1$ is a number determined by the form of the given domain.

1.2 The Analytic Functions of a Complex Variable

If each complex number $z = x + iy$ from some domain G ($G \subset \mathcal{C}$) is associated with the complex number $w = u + iv$, then a function is said to be specified in the domain G:

$$w = w(z) = u(x, y) + iv(x, y).$$

Thus, the specification of a function of a complex variable $w(z)$ is equivalent to the specification of two real functions u, v of two real variables x, y.

Geometrically the function $w(z)$ is a mapping, i.e., the law according to which each point in the domain G is associated with some point in the complex plane w. All the domain of definition G of the function is mapped onto some set of points in the plane w.

Let z_0 be some fixed point in the domain G, and z an arbitrary point in the domain. Let us designate as $\Delta z = z - z_0$ an increment of the independent variable, and $\Delta w = w(z) - w(z_0)$ an increment of the function. The function $w(z)$ is called *continuous at the point* z_0 if $\Delta w \to 0$ as $\Delta z \to 0$. A function is called *continuous in a domain* if it is continuous at each point of a domain.

For the continuity of the function $w(z) = u + iv$ in the domain G it is necessary and sufficient for the real functions $u(x, y)$ and $v(x, y)$ to be continuous in the domain G.

The function $w(z)$ is called *differentiable at the point* z if it has a derivative at this point:

$$w'(z) = \lim_{\Delta z \to 0} \frac{\Delta w}{\Delta z}.$$

The function $w(z)$ is called *differentiable* (or *analytic*) *in the domain* G if it is differentiable at each point of the domain. For the complex function $w = u(x, y) + iv(x, y)$ to be differentiable in a domain it is necessary (but not sufficient!) for the real functions $u(x, y)$ and $v(x, y)$ to be differentiable in this domain.*

The sufficient condition of the differentiability of the function $w = u + iv$ is that the functions u, v must have the continuous partial derivatives u_x, u_y, v_x, v_y, satisfying the *Cauchy-Riemann conditions*

$$u_x = v_y, \quad u_y = -v_x. \tag{1.2.1}$$

The derivative of an analytic function can be presented in any of the four forms:

$$w'(z) = u_x + iv_x = v_y + iv_x = v_y - iu_y = u_x - iu_y.$$

---

*The real function of two variables $u(x, y)$ is called differentiable if it has a differential

$$du = u_x dx + u_y dy.$$

It is known from a course in real analysis that for the function $u(x, y)$ to be differentiable in the domain it is sufficient for its partial derivatives u_x and u_y to be continuous in this domain.

The Cauchy-Riemann conditions (1.2.1) represent the differentiability conditions of the function specified in the Cartesian coordinates. If polar coordinates are used, i.e., if an independent variable is represented in the exponential form $z = \rho e^{i\varphi}$, then for the function $w = u(\rho, \varphi) + iv(\rho, \varphi)$ Cauchy-Riemann conditions (1.2.1) have the following form:

$$\frac{\partial u}{\partial \rho} = \frac{1}{\rho}\frac{\partial v}{\partial \varphi}, \quad \frac{\partial v}{\partial \rho} = -\frac{1}{\rho}\frac{\partial u}{\partial \varphi}. \tag{1.2.2}$$

Alternatively the function itself can be expressed in the exponential form $w = Re^{i\Phi}$. In this case the conditions of the differentiability of a function, specified in the Cartesian coordinates, look like

$$\frac{1}{R}\frac{\partial R}{\partial x} = \frac{\partial \Phi}{\partial y}, \quad \frac{1}{R}\frac{\partial R}{\partial y} = -\frac{\partial \Phi}{\partial x}, \tag{1.2.3}$$

and that of the function specified in polar coordinates, like

$$\rho \frac{\partial R}{\partial \rho} = R \frac{\partial \Phi}{\partial \varphi}, \quad \frac{\partial R}{\partial \varphi} = -R\rho \frac{\partial \Phi}{\partial \rho}. \tag{1.2.4}$$

For analytic functions of a complex variable the rules of differentiation from real analysis are true:

1. The sum and the product of analytic functions $f(z)$ and $g(z)$ are analytic functions and

$$(f + g)' = f' + g',$$
$$(fg)' = f'g + fg'.$$

The quotient of two analytic functions f and g is the analytic function if $g(z) \neq 0$ and

$$(f/g)' = (f'g - fg')/g^2.$$

2. The compound function (the composition of analytic functions) is an analytic function. If the function $w = f(z)$ is analytic in the domain G and maps it onto some domain E, and if the function $W = F(w)$ is analytic in the domain E, then the compound function $W = F(f(z))$ is analytic in the domain G, with

$$W' = F'(w) f'(z).$$

3. The analyticity of the inverse function. Let the function $w = f(z)$ be analytic in the neighborhood of the point z_0 and $f'(z_0) \neq 0$. Then a neighborhood of the point $w_0 = f(z_0)$ can be found in which the inverse function $z = F(w)$ exists satisfying the condition $F(w_0) = z_0$. The inverse function $F(w)$ is analytic and

$$F'(w) = 1/f'(z).$$

The theory of analytic functions is rather a large part of mathematical analysis. We shall mention only main theorems used in the theory of conformal mapping.

THEOREM 1.1
An analytic function has derivatives of any order also being analytic functions.

An integral of a function of a complex variable $w = u + iv$ along the curve L reduces to the sum of two line integrals:

$$\int_L w(z)\, dz = \int_L (u\, dx - v\, dy) + i \int_L (u\, dy + v\, dx).$$

For the analytic functions the Cauchy's theorem is true:

THEOREM 1.2 Cauchy's theorem
If the function $w(z)$ is analytic in the singly connected domain, then the integral along any closed curve lying within this domain equals zero. The integral along the open curve lying in this domain does not depend on the form of the curve but only on the location of its ends.

THEOREM 1.3 The integral Cauchy's formula
If the function $f(z)$ is analytic in the finite domain G with piecewise smooth boundary ∂G and is continuous in the closed domain $\overline{G}$ then at any (internal) point of the domain it is expressed by the integral

$$f(z) = \frac{1}{2\pi i} \int_{\partial G} \frac{f(\xi)}{\xi - z}\, d\xi, \qquad (1.2.5)$$

where the traversal of the boundary of the domain is made in the positive sense, i.e., so that the domain lies to the left.

THEOREM 1.4 Liouville's theorem
If the function $f(z)$ is analytic throughout the whole complex plane C and is bounded then $f(z) = $ const. In other words, no analytic function exists (except for constants) that is bounded throughout the complex plane C.

THEOREM 1.5 Power series and a Taylor's series
For each power series $\sum_{n=0}^{\infty} c_n (z - z_0)^n$ a real (finite or infinite) number $R \geq 0$ can be determined, called the radius of convergence *of the series. If $R = \infty$ then the series converges for any $z \in C$; if $R = 0$, then the series converges only at one point $z = z_0$. If $0 < R < \infty$, then the series converges at $|z - z_0| < R$ and diverges at $|z - z_0| > R$. The domain $|z - z_0| < R$ is called the* circle of convergence *of the power series. The sum of the power series is an analytic function within the circle of convergence.*

The inverse statement is also true: if the function $f(z)$ is analytic in some circle $|z - z_0| < R$, then in this circle it can be expanded into the power series called the Taylor's *series:*

$$f(z) = \sum_{n=0}^{\infty} c_n (z - z_0)^n, \qquad (1.2.6)$$

where $c_0 = f(z_0)$, $c_n = f^{(n)}(z_0)/n!$. The radius of convergence of the Taylor's series equals the distance from the point z_0 to the nearest singular point of the function $f(z)$.

THEOREM 1.6 *A Laurent series*
If the function $f(z)$ is analytic in some annulus $R_1 < |z - z_0| < R_2$ then at each point of this annulus it can be represented as a Laurent *series:*

$$\sum_{n=-\infty}^{\infty} c_n (z - z_0)^n. \qquad (1.2.7)$$

1.2.1 The classification of isolated singular points of single-valued functions

If the function $f(z)$ is analytic in the punctured neighborhood of the point z_0 $(0 < |z - z_0| < R)$, then the point z_0 is called an *isolated singular point* of this function. In the punctured neighborhood under consideration the function $f(z)$ can be expanded into a Laurent series (1.2.7). The classification of isolated singular points is made according to the nature of the Laurent expansion:

1. If the series (1.2.7) does not contain any term with a negative power then the point z_0 is called the *removable* one.

2. If the expansion (1.2.7) contains a finite number of terms with negative powers then the point z_0 is called a em pole. The expansion of a function in the neighborhood of the pole looks like

$$f(z) = \sum_{n=1}^{N} \frac{c_{-n}}{(z - z_0)^n} + \sum_{n=0}^{\infty} c_n (z - z_0)^n,$$

   where $c_{-N} \neq 0$. Integer N is called an order of the pole. The complex number c_{-1} is called the *residue* of the function at the point z_0 and is designated by $c_{-1} = \operatorname{Res} f(z_0)$.

3. If the series (1.2.7) contains the infinite number of terms with negative powers, then the point z_0 is called an *essential singularity*.

Approaching the isolated singular point z_0 an analytic function has:

(a) a finite limit c_0 if the point z_0 is a removable one;

(b) an infinite limit if the point z_0 is a pole;

(c) the function $f(z)$ does not have any limit at the point z_0 if this point is an essential singularity.

Isolated singularities of single-valued functions at the point at infinity can be classified in a similar way. If the function $f(z)$ is analytic outside some circle $|z| > R$ the point at infinity is called an isolated singular point of the function $f(z)$. In the R-neighborhood of an isolated point $z = \infty$ the function $f(z)$ can be expanded into the Laurent series

$$f(z) = \sum_{n=-\infty}^{+\infty} c_n z^n. \tag{1.2.8}$$

The singularities at the point $z = \infty$ are classified according to the structure of the Laurent series:

1. If the series (1.2.8) does not contain any term with a positive power of z then the point $z = \infty$ is called the removable one. In this case the function $f(z)$ has the finite limit c_0 as $z \to \infty$.

2. If the series (1.2.7) has a finite number of terms with positive powers of z then the point $z = \infty$ is called a pole. In this case the expansion of the function in the neighborhood of the point at infinity takes the form

$$f(z) = \sum_{n=-\infty}^{N} c_n z^n,$$

where $c_N \neq 0$. The integer N is called the order of the pole. The complex number $-c_{-1}$ is called the *residue of the function at the point* $z = \infty$ and is designated by $-c_{-1} = \operatorname{Res} f(\infty)$. The function $f(z)$ in this case has the infinite limit as $z \to \infty$.

3. If the series (1.2.8) contains an infinite number of terms with positive powers of z then the point $z = \infty$ is called an essential singularity. In this case the function does not have any limit as $z \to \infty$.

1.2.2   Analytic continuation

The concept of analytic continuation is one of the fundamental ones in the theory of analytic functions. It has no analogies in the theory of functions of a real variable. Let the function $f(z)$ be defined on some set E and the function $F(z)$ be defined and analytic in the domain G that contains the set E. If $F(z) = f(z)$ when $z \in E$ than the function $F(z)$ is called the *analytic continuation* of the function $f(z)$ from the set E into the domain

G. If the set E has at least one limiting point,[†] belonging to the domain G, then the analytic continuation is the unique one.

Sometimes some line is considered as the set E. For example, the elementary functions e^z, $\cos z$, $\ln z$ and so on are defined as the analytic continuations of real functions from the real axis (or half-axis $x > 0$) into the complex plane. However more often a subdomain of the domain G is considered as a set E. In this case the process of analytic continuation allows an extension of the domain of the original specification of a function. The simplest method of an analytic continuation — by power series — was suggested by C. Weierstrass[12] and is called the analytic continuation according to Weierstrass. The initial item of the Weierstrass theory is the concept of an element of an analytic function. An analytic function given in some ε-neighborhood of the given point z_0 is called an element of an analytic function (the neighborhood being able to be as small as possible).

In the ε-neighborhood of the point z_0 the function $f(z)$ can be represented by a power series (compare with (1.2.6)):

$$f(z) = \sum_{n=0}^{\infty} c_n \, (z - z_0)^n,$$

having non-zero radius of convergence R_0.

If the radius of convergence R_0 is infinite then the function

$$F(z) = \sum_{n=0}^{\infty} c_n \, (z - z_0)^n, \qquad (1.2.9)$$

represents the analytic continuation of an element of the analytic function $f(z)$ into the whole complex plane $\mathcal{C}$; if R_0 is finite then $F(z)$ is the analytic continuation of the element $f(z)$ into the circle K_0, defined by the condition $|z - z_0| < R_0$. The analytic function $F(z)$ can be continued outside the boundaries of the circle K_0 by choosing some point z_1 within the circle different from z_0, and constructing the power series

$$F(z) = \sum_{n=0}^{\infty} a_n \, (z - z_1)^n, \qquad (1.2.10)$$

where $a_n = F^{(n)}(z_1)/n!$.

Let the radius of convergence of the series (1.2.10) equal R_1, then the circle of convergence is defined by the condition $|z - z_1| < R_1$. At the points lying inside the circle K_1 the sum of the Taylor's series equals $F(z)$. If it turns out that $R_1 > R_0 - |z_0 - z_1|$ then the circle K_1 would not be entirely located within K_0; in this case the series (1.2.10) represents the analytic continuation $F(z)$ into that part of the circle K_1 which lies outside

---

[†]The *limiting point* of a set is the point that has elements of this set in any punctured neighborhood.

K_0. This process can be continued further by choosing the point z_2 within the circle K_1 and constructing the series:

$$F(z) = \sum_{n=0}^{\infty} b_n\, (z - z_2)^n, \quad b_n = F^{(n)}(z_2)/n!.$$

with the circle of convergence K_2, etc. Thus the analytic continuation is constructed of an element of the analytic function $f(z)$ along the chain of linked circles K_0, K_1, $K_2,\ldots,$.

Considering some smooth line L, starting at the point z_0 and choosing the centers of the circles K_1, $K_2,\ldots$, on this line it is possible to speak about the analytic continuation of an element of the function $f(z)$ along the line L. Here the following theorem is true.

THEOREM 1.7   The monodromy theorem
If in the singly connected domain G containing the point z_0 an analytic continuation of an element $f(z)$ exists along any line starting at the point z_0, then the single-valued function exists in the domain G representing the analytic continuation of the element $f(z)$ into the domain G.

The analytic continuation of the element of the function $f(z)$ along the given line (or along the chain of circles covering it) is defined uniquely. However, in general it is impossible to affirm that starting from the point z_0 and reaching some point z of a plane along different paths we shall receive the same values of the analytically continued functions.

The set of all the analytic continuations of the given element $f(z)$ is called the *complete analytic function*. In general it is multiply-valued and consists of the finite or numbered set of the continuous branches. Since the relation between the two functions, one being the analytic continuation of the other, is reciprocal, the process of analytic continuation is reversible. Hence, the final definition of the complete analytic function does not depend on the point z_0 from which its construction was begun. In the case of a multiply-valued analytic function the branch points are added to the mentioned types of the singular points. The *branch point* of a multiply-valued analytic function is such an isolated singular point that, when traversing around it along a small enough contour, the value of the analytic function changes. In the neighborhood of the branch point it is impossible to rectify a single-valued continuous branch of a function. If after the m-fold traversal of the branch point the function $f(z)$ returns to its value before the traversal, then the branch point is called *algebraic* and the number $(m - 1)$ is called the order of the branch point. For example, for the function $f(z) = (z - a)^{1/m}$‡ the points $z = a$ and $z = \infty$ are branch points of $(m - 1)$ order.

---

‡For the definitions of functions z^{α} and $\ln z$ see further in Section 1.7.

If after any number of traversals of the branch point the function $f(z)$ does not ever return to its value before the traversal, then the branch point is called a *logarithmic* one. For example, for the function $f(z) = \ln(z - a)^{\frac{1}{4}}$ the points $z = a$ and $z = \infty$ are logarithmic branch points.

In this Handbook only single-valued analytic functions and single-valued branches of multiply-valued functions are considered. Single-valued analytic branches can be rectified by making cuts connecting branch points in the domain of definition of the function.

Note one more method of the analytic continuation of a function — the continuation through the boundary of the domain of the initial definition of the function. This method is based on the concept of the regular boundary point of an analytic function.

Let the function $f(z)$ be analytic in some domain G. An unisolated boundary point z_0 in the domain G $(z_0 \in \partial G)$ is called a *regular boundary point* for the function $f(z)$ if the power series

$$F(z) = \sum_{n=0}^{\infty} c_n (z - z_0)^n \qquad (1.2.11)$$

exists having non-zero radius of convergence, its sum being equal to $f(z)$ in the common part of its circle of convergence and the domain G. *Singular boundary points* are the boundary points not being regular.

The function $F(z)$ (1.2.11) is an analytic continuation of the function $f(z)$ from the domain G into that part of a circle of convergence of a power series lying outside the domain G. It is called a continuation of an analytic function through the boundary point z_0. If all the points of some connected part of the boundary γ ($\gamma \subset \partial G$) are regular, then an analytic continuation exists for the function $f(z)$ from the domain G through the part of the boundary γ into the domain, lying outside G. If all the points of the curve γ are singular, then an analytic continuation through the curve γ is obviously impossible. In this case the curve γ is called the *natural boundary* of the function $f(z)$.

An example of a function with a natural boundary is the power series

$$f(z) = \sum_{n=1}^{\infty} z^{n!}. \qquad (1.2.12)$$

The power series (1.2.12) converges in the circle $|z| < 1$. At the points $z = \rho e^{i2\pi m/N}$ it can be represented by

$$f(z) = \sum_{n=1}^{N-1} \rho^n e^{i2\pi mn!/N} + \sum_{n=N}^{\infty} \rho^n ,$$

and hence its sum grows infinitely as $\rho \to 1$. Therefore, all the points $z = e^{i2\pi m/N}$ are singular ones. As these points are placed on the unit circumference $|z| = 1$ densely everywhere, then all the points of the circumference are the singular ones. The circumference $|z| = 1$ is the natural boundary for the function (1.2.12). It is impossible to analytically continue the function (1.2.12) outside the boundary of the unit circle.

In conclusion, we repeat that only one element of the analytical function $f(z)$ given in as small a neighborhood as desired of some point defines completely and uniquely the analytic continuation $F(z)$ — a complete analytic function with all its branches, isolated singularities and the natural boundary.

One more method of an analytic continuation based on the symmetry principle will be considered below in Section 1.6.6.

1.2.3   The classification of analytic functions

The classification of analytic functions is carried out according to the type of all the function singularities. For single-valued functions defined in the complex plane $\mathcal{C}$ (i.e., for functions without natural boundaries) the following classification can be given:

1. Functions that have no singularities in $\overline{\mathcal{C}}$ can only be constants according to Liouville's theorem 1.4.

2. Functions having a singularity only at the point at infinity are called *entire* ones. If the point $z = \infty$ is the pole of the N-order, the entire function represents the polynomial of N-degree:

$$P_N(z) = \sum_{n=0}^{N} c_n\, z^n.$$

   If the point $z = \infty$ is an essential singularity, then the function is called the *entire transcendental* one. For example, such are $\exp(z)$, $\cos(P_N(z))$, $P_N(z)\sin(z)$ and so on.

3. The function having in any finite domain a finite set of poles is called *meromorphic*. If a meromorphic function also has a pole at the point at infinity, then it is a *rational* one:

$$f(z) = P_N(z)/Q_M(z)\,,$$

   where $P_N(z)$, $Q_M(z)$ are polynomials. If the point $z = \infty$ is an essential singularity of the meromorphic function, then it can be represented as a ratio

$$f(z) = \varphi(z)/Q_M(z),$$

   where $\varphi(z)$ is an entire transcendental function, and $Q_M(z)$ is a polynomial. When a meromorphic function has an infinite set of poles, then the point $z = \infty$ is an unisolated singular point (the limiting point of the set of poles). Examples are the functions $1/\sin z$, $\tan(P_N(z))$ and so on. Any meromorphic functions can be represented as a ratio of two entire functions.

We shall not classify functions with essential singularities at finite points of a plane $\mathcal{C}$.

1.3 Conjugate Harmonic Functions

From the Cauchy-Riemann conditions (1.2.1) for the function $w = u + iv$ it follows that the functions $u(x, y)$ and $v(x, y)$ satisfy the equations

$$\frac{\partial u}{\partial x}\frac{\partial v}{\partial x} + \frac{\partial u}{\partial y}\frac{\partial v}{\partial y} = 0, \tag{1.3.1}$$

$$\nabla^2 u = 0, \quad \nabla^2 v = 0, \tag{1.3.2}$$

where $\nabla^2 = \frac{\partial^2}{\partial x^2} + \frac{\partial^2}{\partial y^2}$ is the Laplace operator. Functions satisfying the Laplace equation are called the *harmonic* ones. A pair of the functions $u(x, y)$, $v(x, y)$ connected by the Cauchy-Riemann conditions is called *harmonically conjugate*.

Introducing gradient vectors

$$\nabla u = \mathbf{i}\frac{\partial u}{\partial x} + \mathbf{j}\frac{\partial u}{\partial y}, \quad \nabla v = \mathbf{i}\frac{\partial v}{\partial x} + \mathbf{j}\frac{\partial v}{\partial y},$$

(where $\mathbf{i}$, $\mathbf{j}$ are the unit vectors of a Cartesian coordinate system) equation (1.3.1) can be represented as

$$(\nabla u, \nabla v) = 0, \tag{1.3.3}$$

where parentheses designate the scalar product of two vectors.

Cauchy-Riemann conditions yield

$$|\nabla u| = |\nabla v| = |w'(z)|.$$

Thus, vectors ∇u and ∇v have the same length and are orthogonal, the vector ∇v obtained by the counterclockwise 90^0 rotation of the vector ∇u. Since the vector ∇u is normal to the level lines, defined by the equation $u(x, y) = $ const, the condition (1.3.3) can be interpreted as a condition of mutual orthogonality of level lines families, defined by the equations

$$u(x, y) = \text{const} \quad \text{and} \quad v(x, y) = \text{const}. \tag{1.3.4}$$

A set of level lines of a pair of conjugate harmonic functions is said to be an *isothermic net*. Therefore, an arbitrary analytic function $w(z)$ determines an isothermic net of lines which are given by equations (1.3.4). At the points where $w'(z) = 0$, the vectors ∇u and ∇v vanish and the direction of the isothermic net lines is uncertain.

Introducing a new variable $U = \ln R$ one can transform Cauchy-Riemann conditions (1.2.3) for the function $w = Re^{i\Phi}$:

$$\frac{\partial U}{\partial x} = \frac{\partial \Phi}{\partial y}, \quad \frac{\partial U}{\partial y} = -\frac{\partial \Phi}{\partial x}$$

(it is assumed that $w \neq 0$, i.e., $R > 0$). Therefore, the functions U and Φ are harmonically conjugate and

$$(\nabla U, \nabla \Phi) = 0, \quad |\nabla U| = |\nabla \Phi|.$$

Since $\nabla U = \frac{1}{R} \nabla R$, then

$$(\nabla R, \nabla \Phi) = 0, \quad \frac{1}{R} |\nabla R| = |\nabla \Phi|. \tag{1.3.5}$$

The vectors ∇R and $\nabla \Phi$ represent normals to level lines of the functions $R(x, y)$, $\Phi(x, y)$, i.e., to lines determined by conditions

$$R(x, y) = \text{const}, \quad \Phi(x, y) = \text{const}. \tag{1.3.6}$$

Because of the condition (1.3.5) families of lines (1.3.6) are reciprocally orthogonal. These two families form *the relief map of analytic function* $w = f(z)$.*

If in a singly connected domain G one of the conjugate harmonic functions is given ($u(x, y)$, for example), then another function ($v(x, y)$) can be found within an arbitrary constant term. Indeed, let us consider a real line integral

$$V(x, y) = \int_{M_0 LM} (-u_y \, dx + u_x \, dy) + C \tag{1.3.7}$$

where L is an arbitrary curve having a fixed initial point $M_0(x_0, y_0)$ and terminal point $M(x, y)$ and belonging entirely to the domain G; $C = V(x_0, y_0)$

---

*The surface in three-dimensional space (x, y, Z) determined by the equation $Z = |f(x + iy)|$ is called the *relief* of the function of a complex variable $w = f(z)$. Sections of this surface by the planes $Z = $ const are called *horizontals of the relief*, projections of horizontals onto the plane (x, y) form the *map of horizontals* of the relief of the function $f(z)$. The lines, directed along the vector $\nabla R = \nabla |f(z)|$, i.e., being orthogonal to the horizontals of the relief map, are called steepest descent curves. In the case of the analytic function $f(z)$ descent curves coincide with the lines $\arg f(z) = $ const. The direction of horizontals and descent curves of the function $f(z)$ is determined at all points where $f(z) \neq 0$, $f'(z) \neq 0$.

is an arbitrary constant. The integral (1.3.7) does not depend on the integrating path L.[†]

The differential of the function (1.3.7) is equal to

$$dV = -u_y\, dx + u_x\, dy,$$

or in other words,

$$V_x = -u_y, \quad V_y = u_x.$$

Therefore, the function (1.3.7) is the conjugate harmonic to $u(x, y)$ and it is determined within the constant C. If a real (or an imaginary) part of some analytic function is given in some singly connected domain, then the function itself can be reconstructed within a constant term. The solution of the problem exists only when a given function is a harmonic one. In a multiply connected domain, the problem under investigation can have a multiply-valued (infinitely-valued) analytic function as a solution.

1.4 The Geometric Meaning of the Derivative. Isogonal and Local Conformal Mappings

The function $w = f(z)$, defined in some domain G $(G \subset C)$, maps this domain onto some set of points E of the complex plane w. The set E is called *image* and the domain G *preimage* in the mapping, which is carried out by $w = f(z)$ function. Analytic functions possess a *property of domain conservation*: if the function $f(z)$ is analytic in G and $f(z) \neq$ const, then the set E is a domain. Generally speaking non-analytic functions do not possess such a property. For example, the functions

$$w_1 = z + \overline{z}, \quad w_2 = z\overline{z}$$

map any domain into segments of a real axis.

Analytic functions have an analogous property concerning lines: any curve, lying in the domain G, is mapped by the analytic function $w = f(z)$

[†]From a course in real analysis it is known that for the line integral

$$\int_L P(x, y)\, dx + Q(x, y)\, dy,$$

to be independent of the shape of the curve, where P and Q are continuous and differentiable functions in the singly connected domain G, containing the curve L, it is necessary and sufficient for the functions P and Q to satisfy the condition

$$\frac{\partial P}{\partial y} \equiv \frac{\partial Q}{\partial x}.$$

into a curve, disposed in the domain E. An image of the curve can be not a curve for non-analytic functions. For example, for the function $w_1 = z + \bar{z}$ an image of the vertical line $\operatorname{Re} z = c$ is the point $w_1 = 2c$ and for function $w_2 = z\bar{z}$ an image of the circumference $|z| = R$ is the point $w_2 = R^2$.

Let the function $w = f(z)$ be analytic at the point $z_0{}^*$ ($z_0 \subset C$) and $f'(z_0) \neq 0$. Further let γ be an arbitrary smooth Jordan curve with the initial point z_0:

$$z = z(t), \quad t_0 \leq t \leq t_1, \quad z(t_0) = z_0.$$

An image of the curve γ in the mapping $w = f(z)$ is the smooth curve Γ with the starting point $w_0 = f(z_0)$. The directions of tangents to the curves γ and Γ at the points z_0 and w_0, respectively, are determined by the linear elements

$$dz = z'(t_0)\, dt, \quad dw = f'(z_0)\, z'(t_0)\, dt = f'(z_0)\, dz.$$

Designating $f'(z_0) = k\, e^{i\alpha}$, a relation between linear elements of the curves γ and Γ can be expressed by the equality

$$|dw| = k\,|dz|, \quad \arg(dw) = \arg(dz) + \alpha.$$

The value $k = |f'(z_0)|$ represents a *stretch coefficient* of a curve element at the point z_0 in the mapping $w = f(z)$ and the value $\alpha = \arg f'(z_0)$ — an *angle of rotation* of the curve γ element in that mapping.

If γ_1 and γ_2 are two smooth curves intersecting at the point z_0 with the angle φ between them, then the angle between their images in the mapping $w = f(z)$ also is equal to φ. Such mappings that preserve absolute values of angles between curves as well as the directions of the angles are called *isogonal* ones.

If $f'(z_0) = 0$ then the mapping $w = f(z)$ is not isogonal at the point z_0. Therefore, the mapping carried out by the analytic function $w = f(z)$ is isogonal at the point z_0 ($z_0 \in C$) if and only if $f'(z_0) \neq 0$.

A concept of isogonal mapping can be extended to the cases in which one of the points z_0 or w_0 (and even both) is the point at infinity. In these cases as the angle between two smooth curves which pass through the point $z = \infty$, one can consider the angle between images of these curves at the point $\zeta = 0$ in the mapping $\zeta = 1/z$. For example, if curves γ_1 and γ_2 have asymptotes with the angle φ at the finite point of their intersection, then the angle between curves γ_1 and γ_2 at the point ∞ equals $-\varphi$ in accordance with this definition.

---

*The function $f(z)$ is called *analytic at the point* $z_0 \neq \infty$ if it is analytic in some neighborhood of the point z_0. The function $f(z)$ is called *analytic at the point* $z_0 = \infty$ if the function $F(\zeta) = f(1/\zeta)$ is analytic at the point $\zeta = 0$.

Conditions of mapping isogonality at the point $z_0 = \infty$ are as follows. Let a function $w = f(z)$ be analytic at the point $z_0 = \infty$, i.e., the function

$$F(\zeta) = f(1/\zeta)$$

be analytic at the point $\zeta = 0$. The function $F(\zeta)$ can be expanded into a power series in a neighborhood of the point $\zeta = 0$

$$F(\zeta) = \sum_{n=0}^{\infty} c_n\, \zeta^n. \tag{1.4.1}$$

The mapping (1.4.1) is isogonal at the point $\zeta = 0$, if $F'(0) = c_1 \neq 0$. The following Laurent series for the function $f(z)$ corresponds to the expansion (1.4.1)

$$f(z) = \sum_{n=0}^{\infty} \frac{c_n}{z^n}.$$

This mapping is isogonal at the point $z = \infty$ if and only if $c_1 \neq 0$, i.e., when the residue of the function $f(z)$ with respect to the point at infinity does not vanish: $\operatorname{Res} f(\infty) \neq 0$.

Conditions of mapping isogonality at the point $w_0 = \infty$. The analytic function $w = f(z)$ maps some point z_0 ($z_0 \in C$) into a point $w_0 = \infty$ if and only if the point z_0 is a pole. If the order of the pole equals N, then the function $f(z)$ can be represented as

$$f(z) = \frac{\varphi(z)}{(z - z_0)^N},$$

where $\varphi(z)$ is an analytic function and $\varphi(z_0) \neq 0$. The mapping carried out by the function $w = f(z)$ is isogonal at the point z_0, if the mapping

$$\zeta = \frac{1}{w} = \frac{(z - z_0)^N}{\varphi(z)}$$

is isogonal at this point. This last mapping satisfies the isogonality condition $\zeta'(z_0) \neq 0$ if and only if $N = 1$, i.e., if the point z_0 is a pole of the first order.

This statement is also true in the case when $z_0 = \infty$ and $w_0 = \infty$. Indeed, in this case the function $w = f(z)$ has a pole at the point $z_0 = \infty$ and can be represented as

$$f(z) = z^N \varphi(1/z),$$

where N is the order of the pole, $\varphi(\zeta)$ is the function analytic at the point $\zeta = 0$ and $\varphi(0) \neq 0$. Isogonality of the mapping $w = f(z)$ at the point

$z_0 = \infty$ means isogonality of the mapping

$$W = \frac{1}{f(1/\zeta)} = \frac{\zeta^N}{\varphi(\zeta)}$$

at the point $\zeta = 0$. The derivative $dW/d\zeta$ does not vanish as $\zeta = 0$ if and only if $N = 1$. Therefore, the mapping $w = f(z)$, which transforms some point z_0 (finite or the point at infinity) into the point $w_0 = \infty$, is isogonal at the point z_0 if and only if the point z_0 is a pole of the first order.

If the function $w = f(z)$ is analytic at the point z_0 and $f'(z_0) \neq 0$ then the mapping realized by it preserves angles between curves as well as the shape of infinitesimal figures lying in the neighborhood of the point z_0. Indeed, the differentiability of the function $f(z)$ yields

$$\Delta w = f'(z_0)\,\Delta z + \varepsilon\,\Delta z,$$

where $\Delta z = z - z_0$, $\Delta w = w(z) - w(z_0)$ and $\varepsilon \to 0$ as $\Delta z \to 0$. Considering Δz as a segment of the line connecting the points z_0 and $z = z_0 + \Delta z$ we obtain that it is mapped into some curve by the mapping $w = f(z)$. Its chord up to the second-order term is equal to

$$\Delta w = f'(z_0) \cdot \Delta z.$$

In this mapping an infinitesimal triangle formed by two vectors Δz_1 and Δz_2 is transformed into the similar infinitesimal triangle with sides Δw_1 and Δw_2. Linear sizes of it are magnified by factor of $k = |f'(z_0)|$ while the area is magnified by $k^2 = |f'(z_0)|^2$ times. Any infinitesimal figure is mapped by the function into a similar one.

A geometrical transformation, in which any infinitesimal figure belonging to the neighborhood of the point z_0 is mapped into the similar one is called a conformal mapping at the point z_0 or in other words a *locally conformal* one. For the mapping $w = f(z_0)$ to be isogonal and conformal at the point z_0 it is necessary and sufficient that the function $f(z)$ be analytic at the point z_0 and satisfy the condition $f'(z_0) \neq 0$.

Sometimes the concept of *conformal mapping of the second class* is used. This is the term for the mapping preserving the absolute values of the angles between curves, but changing their direction. Such mappings are realized by the functions $w = \overline{f(z)}$, where $f(z)$ is an analytic function. They can be considered a sequence of a conformal mapping $\zeta = f(z)$ and a mirror reflection $w = \overline{\zeta}$ about the real axis of the ζ-plane.

Mappings carried out by non-analytic functions are not conformal ones.

Exercise

1. With the help of the program CONFORM create on the computer screen the mappings of Cartesian and polar nets carried out by the non-analytic function

$$w = z + 0.3 \cdot i \cdot \overline{z}.$$

    Show that the mapping is not conformal: infinitesimal circles are mapped by this function into ellipses, and squares into parallelograms.

1.5 Univalent Analytic Functions. Conformal Mappings of Domains

A function $f(z)$ defined in a domain G is called a univalent function in this domain if at distinct points of the domain it assumes distinct values, i.e., if the condition $z_1 \neq z_2$ ($z_1, z_2 \in G$) yields $f(z_1) \neq f(z_2)$. Univalence of the function means the inverse function is single-valued. The mapping carried out by an univalent function is reciprocal one-to-one.

For an analytic function $f(z)$ the necessary condition of its univalence in the domain G is that $f'(z) \neq 0$ in this domain. For a small enough neighborhood of the point z_0 this condition is also a sufficient one, as it follows from the theorem of the existence and uniqueness of the inverse function (see Section 1.2). In general, the condition $f'(z) \neq 0$ is not sufficient for the univalence. For example, in the half-plane Re $z > 0$ the function $f(z) = z^4$ is not univalent because it assumes the same values at distinct points of the domain (for example, at the points $z = e^{i\pi/4}$ and $z = e^{-i\pi/4}$ of the half-plane the function is equal to -1). However, the derivative of the function $f'(z) = 4z^3$ in the half-plane Re $z > 0$ does not vanish anywhere.

The zeros of the derivative of the function $f(z)$, i.e., the points where $f'(z) = 0$, are called *critical points* of the function $f(z)$. In the domain containing critical points the function is not univalent.

Let the function $w = f(z)$ defined in some domain G ($G \subset \mathcal{C}$) map it onto another domain D. Such a mapping of one domain onto another is called conformal if it is reciprocal one-to-one and isogonal at each point of the domain G. For the mapping of the domain G realized by the function $w = f(z)$ to be conformal it is necessary for the function $f(z)$ to be univalent and analytic in G besides, probably, one point at which the function may have the first order pole. At all the finite points of analyticity the function $f(z)$ must satisfy the condition $f'(z) \neq 0$, at the point at infinity — the condition Res $f(\infty) \neq 0$.

The definition of a conformal mapping together with the simplest properties of analytic functions (see Section 1.2) yield the basic properties of conformal mappings:

1. A mapping inverse to the conformal one is also conformal. In other words, if the function $w = f(z)$ maps the domain G conformally onto the domain D, then the inverse function $z = F(w)$ maps the domain D conformally onto the domain G.

2. The composition of conformal mappings represents a conformal mapping. In other words, if the function $w = f(z)$ maps the domain G conformally onto the domain D and the function $W = F(w)$ maps the domain D conformally onto the domain E, then the compound function $W = F(f(z))$ maps the domain G conformally onto the domain E.

A conformal mapping represents the particular, special case of the general substitution of variables (or of the coordinate transform), considered in real analyses:

$$u = u(x, y), \quad v = v(x, y). \tag{1.5.1}$$

In the case of a conformal mapping variables (1.5.1) are connected by the Cauchy-Riemann conditions (1.2.1). The Jacobian of this change of variables equals

$$\begin{vmatrix} u_x & u_y \\ v_x & v_y \end{vmatrix} = u_x^2 + v_x^2 = |f'(z)|^2 > 0.$$

Let the function $w = u + iv = f(z)$ map the domain G conformally onto the domain D. If the domain G is finite and the function $f(z)$ in this domain is limited, then the domain D is also finite. The area of the domain D is expressed in the form of a double integral

$$S = \iint_G |f'(z)|^2 \, dx \, dy.$$

The transformation of differential expressions with conformal transform of variables (1.5.1) will be considered in Section 1.8.

The variables u and v which uniquely determine the location of a point z in the domain G represent curvilinear coordinates of the point z. These coordinates are called *conformal* or *isothermic* ones.

Therefore, conformal coordinates are a particular case of orthogonal curvilinear coordinates in the plane, and are themselves a particular case of the general change of variables (1.5.1). Particular cases of conformal coordinates in the plane — parabolic, elliptic and bipolar ones — will be considered further in Section 1.7.

Let u, v be the conformal coordinates defined by the function $w = u + iv = f(z)$. Coordinate lines of this curvilinear coordinate system determined by the equations

$$u(x,y) = \text{const}, \quad v(x,y) = \text{const},$$

form the isothermic net of the function $f(z)$. This net is the image of the Cartesian net in the domain D in the inverse conformal mapping $z = F(w)$ (where $F = f^{-1}$ is the inverse function).

The image of a polar net of lines in the domain D in the inverse conformal mapping $z = F(w)$ is represented by a net of lines defined by equations

$$|w| = |f(z)| = \text{const}, \quad \arg w = \arg f(z) = \text{const},$$

i.e., the relief map of the function $f(z)$ (see Section 1.3).

___

Exercises

1. To create with the help of the program CONFORM the mapping of a polar net in the circle $|z| < R$ realized by a function

$$w = z + z^2. \tag{1.5.2}$$

   To choose experimentally the maximum value of R when the function is still univalent inside the circle. To examine whether the only critical point ($z = -1/2$) lies on the boundary of the maximum univalence circle.

   Hint. The violation of the univalence can be detected by overlapping of different parts of the mapping.

2. To do the same with the function

$$w = \sum_{n=1}^{\infty} z^{n!} = z + z^2 + z^6 + z^{24} + \dots.$$

   As $R < 1/2$, then in the series under consideration one can limit oneself by the first three terms.

3. To do the same with the function

$$w = \sum_{n=0}^{\infty} z^{2^n} = z + z^2 + z^4 + z^8 + z^{16} + \dots.$$

   This series, like (1.2.12), has the natural boundary $|z| = 1$. The radius of the univalence circle $R < 1/2$. Inside the univalence circle one can limit oneself by 4 terms.

___

1.6 General Principles of the Theory of Conformal Mappings

The main problem of the theory of conformal mappings is to find an analytic function carrying out the mapping of the given domain onto another given domain. As a universal algorithm for the solution of the problem does not exist, more important are the general principles of conformal mappings used when solving concrete problems.

1.6.1   The principle of domain preservation

The principle of domain preservation true for arbitrary analytic functions (see Section 1.4) acquires supplementary features in the case of conformal mappings. Formerly, in the conformal mapping of domains the order on connectedness of a domain is preserved, i.e., a singly connected domain is mapped conformally into a singly connected one; doubly connected domain into a doubly connected one, etc.

Let the function $w = f(z)$ analytic in the domain G and continuous in the closed domain $\overline{G}$ map the domain G conformally onto some domain D $(D = f(G))$. At that the boundary of the domain G is mapped onto the boundary ∂D. Isolated boundary points of the domain G are mapped into that of the domain D, and nondegenerate parts of the boundary of one domain into that of the other one. If the boundaries of the domains G and D represent Jordan curves or if they consist of N non-crossing Jordan curves then the function $w = f(z)$ lays down reciprocal one-to-one and continuous correspondence (homeomorphism) between points of the boundaries ∂G and ∂D. In general, the mapping of the boundary ∂G into ∂D is not reciprocal one-to-one, because multi-fold points of the boundary ∂G can correspond with one-fold points of the boundary ∂D and vice versa. However, introducing the concept of boundary traversal line and considering all its points to be different, the function $w = f(z)$ in a conformal mapping can be affirmed to map the boundary ∂G onto the boundary ∂D reciprocally one-to-one with the orientation preserved.*

According to the definition a conformal mapping is isogonal at each of interior points of a domain. Generally speaking, at boundary points of a domain the mapping $w = f(z)$ is not isogonal. However, at regular boundary points of the domain G a mapping is isogonal. Local conformity can be disturbed at singular points of the boundary ∂G only.

1.6.2   Theorems of existence

For singly connected domains the theorem of existence of a conformal mapping belongs to B. Riemann. Riemann's theorem states that for any connected domain of a complex z-plane with a nondegenerate boundary an analytic function $w = f(z)$ exists mapping this domain conformally onto

---

*The boundary of a domain is called oriented if its traversal direction is defined. The positive sense of traversal of the nondegenerate boundary part we call the sense for which the domain is always to the left.

the unit circle $|w| < 1$. The corollary from this theorem yields that any non-degenerate singly connected domain can be mapped conformally onto any nondegenerate singly connected domain by a countless number of means. Riemann's theorem represents a typical theorem of existence. It does not say anything about how to find a mapping function.

Riemann's theorem allows us to divide all the singly connected domains into three classes:

1. The extended complex plane $\overline{C}$.

2. The extended complex plane $\overline{C}$ with one point deleted (finite or the point at infinity).

3. Nondegenerate singly connected domains.

The extended complex plane $\overline{C}$ can be conformally mapped onto the extended complex plane $\overline{C}$ only. The mapping is carried out by linear-fractional functions (see below Subsection 1.7.2).

The extended complex plane $\overline{C}$ with one point deleted can be conformally mapped only onto the extended complex plane $\overline{C}$ with one point deleted, for example, onto the open complex plane C. Such a mapping can be realized also by linear-fractional functions. A nondegenerate singly connected domain according to Riemann's theorem can be conformally mapped onto any nondegenerate singly connected domain, for example, onto a circle, a half-plane, a strip, etc.

For multiply connected domains theorems of existence of conformal mappings are much more complicated. Already for doubly connected domains it is impossible to state the possibility of a conformal mapping of a given domain onto any doubly connected domain. However, one can affirm that for the given doubly connected domain with nondegenerate boundaries such an annulus $1 < |w| < R$ can be found, onto which the given domain can be mapped conformally. The number R is called the modulus of a doubly connected domain. It can be uniquely defined by the form of a given domain. All the doubly connected domains with the same R can be mapped conformally onto each other.

The N-connected domain of the complex z-plane can be mapped conformally onto the w-plane with N non-crossing circles deleted (Koebe's theorem) as well as onto the plane with N cuts along the parallel straight line segments (Hilbert's theorem). The sizes and locations of the circles (or segments) are not arbitrary but are defined by the form of the given N-connected domain.

1.6.3 Theorems of uniqueness

Theorems of uniqueness of a conformal mapping will be formulated only for singly connected domains.

Riemann's theorem states the existence of the function $w = f(z)$, realizing the conformal mapping of a given nondegenerate singly connected domain onto a unit circle $|w| < 1$. The function $f(z)$ is defined uniquely if some interior point z_0 of the domain is required to be mapped into the center of the circle and the angle of a turn of a linear element at the point z_0 is required to equal the given number α:

$$f(z_0) = 0, \quad \arg f'(z_0) = \alpha.$$

In the general case Riemann's theorem claims the existence of the function $w = f(z)$ realizing the conformal mapping of nondegenerate singly connected domain G onto nondegenerate singly connected domain D. The conditions supplying the uniqueness of such a mapping are called the normalization conditions. The most useful are three following forms of normalization conditions:

1. $f(z_0) = w_0$, $\arg f'(z_0) = \alpha$, where z_0, w_0 are arbitrary interior points of the domains G and D correspondingly, α is an arbitrary real number;

2. $f(z_0) = w_0$, $f(z_1) = w_1$, where z_0, w_0 are interior, z_1, w_1 are boundary points of domains G and D correspondingly;

3. $f(z_1) = w_1$, $f(z_2) = w_2$, $f(z_3) = w_3$, where z_1, z_2, z_3 are three different boundary points of the domain G, numbered in the order of the positive sense of traversal of the boundary ∂G; w_1, w_2, w_3 are different boundary points of the domain D, numbered in the order of the positive traversal direction of the boundary ∂D.

The theorem of the existence and uniqueness of a conformal mapping for singly connected domains can be formulated as follows.

THEOREM 1.8   *Theorem of Existence and Uniqueness*
If G is a nondegenerate singly connected domain of the z-plane and D is a nondegenerate singly connected domain of the w-plane then the unique function $w = f(z)$ exists, which realizes the conformal mapping of the domain G onto the domain D and satisfies the normalization conditions 1–3.

1.6.4   The principle of univalence

For the mapping carried out by the function $w = f(z)$ analytic in the domain G to be conformal, it is necessary by definition for the mapping to be reciprocal one-to-one, i.e., the function $f(z)$ must be univalent in the domain G. The univalence condition turns to be as well the sufficient condition of the conformality of a mapping, as it is formulated in the following theorem:

THEOREM 1.9   *The Principle of Univalence*

The mapping realized by the function $w = f(z)$ analytic in the domain G is conformal when and only when the function $f(z)$ is univalent in the domain G.

Formulated in general form, this principle becomes the criterion of the conformality of a mapping.

THEOREM 1.10   *The Conformality Criterion*

The function $w = f(z)$ given in the domain G maps this domain conformally onto the domain $D = f(G)$ when and only when this function is univalent and analytic in G everywhere except for, possibly, one point at which the function can have a pole.

1.6.5   The principle of correspondence of boundaries

Conformal mappings, by definition, are reciprocal one-to-one, i.e., they are realized by univalent functions. However, in many cases it is very difficult to check the univalence of a function in a given domain, while to examine the one-to-one correspondence of the domain boundary in the mapping is much easier. To test the conformality of the mapping of one domain onto another it is often sufficient to examine the mapping of the boundaries of these domains. Theorems allowing one to judge the univalence of the function in the domain according to the mapping of the boundary of this domain are united under the name of the principle of correspondence of boundaries. We formulate them only for singly connected domains and first for that with finite boundaries.

THEOREM 1.11

Let G be a singly connected domain (finite or infinite) with the boundary ∂G being a closed (finite) Jordan curve; let $w = f(z)$ be a function analytic in the domain G and continuous in the closed domain $\overline{G}$ mapping one-to-one reciprocally the boundary ∂G onto a closed Jordan curve Γ. Then the function $f(z)$ maps the domain G conformally onto the domain D_i lying inside the curve Γ. At that the orientation of the boundary does not change, i.e., for the positive sense of the boundary traversal ∂F the point $w = f(z)$ traverses the curve Γ in the counterclockwise direction.

THEOREM 1.12

Let G be a singly connected domain (finite or infinite) with the boundary ∂G being a closed Jordan curve; let $w = f(z)$ be an analytic function in the domain G everywhere except for one point — the first order pole, and continuous in $\overline{G}$.[†] *If the function $f(z)$ maps the boundary ∂G one-to-one reciprocally onto a closed Jordan curve Γ, then it maps conformally the domain G onto the domain D, external with respect to contour Γ. At that the orientation of the boundary is preserved, i.e., the positive sense of the boundary traversal ∂G corresponds to the traversal of the curve Γ in the clockwise direction.*

We will usually consider some canonical singly connected domain (a circle, an exterior of a circle, a half-plane, a strip) as the domain G of the function definition. In these cases the boundary ∂G represents an ideally smooth Jordan curve. Nevertheless the image of the boundary ∂G in the mapping $w = f(z)$, i.e., the curve Γ can be a non-Jordan curve. In order to extend the Theorems 1.11, 1.12 to the case of domains with multi-fold boundary points it is necessary to at first agree how to understand reciprocal one-to-one correspondence of the boundaries.

In these cases there is usually introduced a concept of a traversal line of the domain boundary as a continuous closed curve lying inside the domain infinitely close to its boundary, and all the points of such a line being assumed different. At that, each of the multi-fold points of the domain boundary is counted as many times as its multiplicity is equal to. Such a procedure allows us to establish a reciprocal one-to-one correspondence of domain boundaries.

In the case when the domain G is a circle with boundary mapped onto a finite non-Jordan curve, the Theorem 1.11 can be reformulated as follows:

THEOREM 1.13
If the function $w = f(z)$ analytic in the circle $|z| < 1$ and continuous in the closed circle $|z| \leq 1$ maps the circumference $|z| = 1$ reciprocally one-to-one and continuously onto the traversal line of the boundary of some finite singly connected domain D with preservation of orientation, then the function is univalent in the unit circle and maps it conformally onto the domain D.

It should be specially mentioned that the demand of preservation of the traversal orientation of the boundary has been added to the conditions of the theorem. Simple examples show that in the cases when the image of the boundary is a non-Jordan curve, the demand can not be excluded. For

---

[†] Here and further the continuity of a function is accounted for in a generalized sense: the function $f(z)$ is called continuous at the point z_0 if $f(z_0) = \lim\limits_{z \to z_0} f(z)$ and the opportunity of $f(z_0) = \infty$ (as well as the opportunity of $z_0 = \infty$) is not excluded. Therefore if the point z_0 is a pole of the function $f(z)$ and $f(z_0) = \infty$, then the function $f(z)$ is continuous at the point z_0.

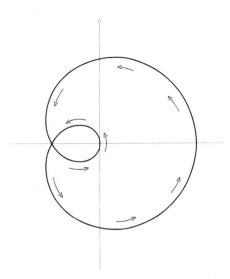

FIGURE 1.5

example, the function

$$w = z + z^2 \qquad (1.6.1)$$

analytic inside the circle $|z| < 1$ maps the circumference $|z| = 1$ onto the closed curve with self-intersection (Pascal's limaçon, see Section 1.10.4 below) being the boundary of some singly connected domain enclosed between the loops of the curve. The mapping of the circumference onto the Pascal's limaçon is reciprocal one-to-one everywhere except for the point of self-intersection. However the function (1.6.1) is not univalent in the circle $|z| < 1$. Nonconformity of the mapping here is a consequence of the condition of the traversal direction preservation having been violated at the interior loop of the curve (Fig. 1.5).

The demand of boundary orientation preservation is essential in the mapping of domains with infinite non-Jordan boundaries as well. For example, the function $w = \sin z$ analytic in the strip $-\pi/4 < \mathrm{Re}\, z < 3\pi/4$ maps reciprocally one-to-one the boundary straight lines of the strip onto two branches of equi-sided hyperbola $u^2 - v^2 = 1/2$, which forms the boundary of the singly connected domain (curvilinear strip) enclosed inside. However under the positive sense of the boundary traversal of the strip $-\pi/4 < \mathrm{Re}\, z < 3\pi/4$ the boundary of the curvilinear strip is not traversed in the positive sense (Fig. 1.5). The condition of traversal direction preservation is violated for the right branch of the hyperbola. As a result the mapping of the mentioned strip by the function $w = \sin z$ is nonconformal.

The principle of correspondence of boundaries is formulated somewhat more complicatedly for the domains with infinite boundaries. Simple examples show that the theorem 1.11 can not be extended on this case without additional propositions. For example, the functions

$$w = z^3, \quad w = \sinh z$$

analytic in the whole complex plane $\mathcal{C}$ map the real axis Im $z = 0$ onto the real axis Im $w = 0$ in a one-to-one manner. However these functions are neither univalent in the lower half-plane nor in the upper one.

Let us formulate the principle of correspondence of boundaries for the case when the function $w = f(z)$ maps an infinite straight line onto an infinite Jordan curve.

THEOREM 1.14
Let the function $w = f(z)$ analytic in the half-plane Im $z > 0$ and continuous in the closed domain Im $z \geq 0$ grow as $|z| \to \infty$, Im $z > 0$ not faster than $C z^2$.[‡] If the function maps the real axis Im $z = 0$ reciprocally one-to-one onto some infinite Jordan curve Γ, then the function is univalent in the upper half-plane and maps it conformally onto curvilinear angular domain lying to the left of the curve Γ.[§]

The next and the last of the theorems about the correspondence of boundaries concern the case when the point at infinity is a K-fold point of the curve Γ. Let us take the circle $|z| < 1$ as the domain of $w = f(z)$ function definition. Assume that the circumference $|z| = 1$ is mapped onto the curve Γ being the boundary of some singly connected domain D (curvilinear strip with K branches). Let denote as $a_1, a_2, \ldots, a_K$ the points of a unit circumference where the function $f(z)$ is infinite (these points are preimages of strip branches). Assume that at every point a_k the limit

$$\lim_{z \to a_k} \left[f(z) \left(z - a_k \right)^{p_k} \right] = c_k \neq 0$$

exists, where p_k is a real positive number. Otherwise this condition can be written as

$$f(z) \sim \frac{c_k}{(z - a_k)^{p_k}}.$$

The number p_k is called an order of growth of the function $f(z)$ at the point $z = a_k$. Under these designations the theorem on the correspondence of boundaries for the curvilinear strip with K branches can be formulated as:

[‡]It means that the ratio $f(z)/z^2$ is bounded as $0 \leq \arg z \leq \pi$ for any sufficiently large $|z|$.

[§]The traversal sense of the curve Γ corresponds to positive direction on the straight line Im $z = 0$.

THEOREM 1.15
Let the function $f(z)$ analytic in the circle $|z| < 1$ and continuous (in generalized sense) in the closed circle $|z| \leq 1$ map reciprocally one-to-one and with preservation of orientation the circumference $|z| = 1$ onto K non-intersecting infinite Jordan curves, which forms the boundary of some singly connected domain D, the points a_1, a_2,...,a_k are mapped into the point at infinity. If the function $f(z)$ has the order of growth p_k ($p_k > 0$), i.e., $f(z) \sim C_k/(z - a_k)^{p_k}$ as $z \to a_k$ and $\sum_{k=1}^{K} p_k \leq 2$, then the function $f(z)$ is univalent inside the circle $|z| < 1$ and conformally maps it onto the domain D.

The theorems 1.14–1.15 can be extended for the case of non-Jordan boundaries if, as above, to introduce a concept of the traversal line of the boundary of a domain which has multiple boundary points.

Applications of the principle of correspondence of boundaries to the construction of some mappings are considered further in Sections 1.10 and 1.11.

Exercises

1. With the help of the program CONFORM create the mapping of a Cartesian net in the upper half-plane Im $z > 0$ carried out by the function

$$w = z^p$$

for various values $p > 0$. Make sure that if $p \leq 2$ the function z^p is univalent in the given half-plane and is not univalent if $p > 2$.

Hint. It is recommended to create the mapping of the rectangle

$$-a < x < a, \quad 0.001 < y < a,$$

choosing the parameter a and the scaling factor experimentally.

2. Make the same for the function

$$w = (z + i)^p,$$

where $p > 0$. Explain the result with the help of the theorem 1.14.

3. Create the mapping of the polar net in the unit circle carried out by the function

$$w = \frac{z}{(1 + z^k)^p},$$

where k is a natural number, $p > 0$ is a real one. Make sure that if $p \leq 2/k$ the mapping is univalent and for $p > 2/k$ it is not univalent. Explain the result with the help of the theorem 1.14.

1.6.6   Symmetry principle

The so-called symmetry principle is widely used when solving concrete problems on the conformal mappings of domains having the axis of sym-

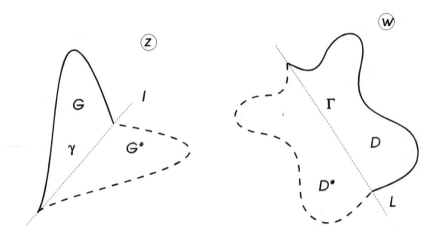

FIGURE 1.6

metry. This principle of the theory of conformal mappings was first formulated by B. Riemann[13] and proved by H. A. Schwarz,[21] so it is often called
the Riemann-Schwarz symmetry principle.

   In the simplest (and the most commonly used) form of the symmetry
principle one considers some finite or infinite domain D with a boundary
having a straight line part γ. A segment or a ray of some straight line l or
the whole infinite straight line l can be such a part.

   The domain G^* obtained by means of a mirror reflection of the domain
G about the straight line l is called symmetrical to the domain G with
respect to the straight line l. The domain G is assumed to lie to one side
of the straight line l and the domains G and G^* do not intersect.

   Let the function $w = f(z)$ analytic in the domain G and continuous in
the closed domain $\overline{G}$ map G conformally onto some domain D with the
boundary also having a straight line part Γ belonging to the straight line
L. Let the function $w = f(z)$ map the straight line part γ onto the straight
line part Γ, i.e., $\Gamma = f(\gamma)$. The parts γ and Γ are depicted in Fig. 1.6 by
dotted lines.

   On the assumptions the following theorem is valid:

THEOREM 1.16   The symmetry principle
Let the function $f(z)$ be analytic in the domain G, the boundary of which
includes the straight line part γ, and continuous in the closed domain $\overline{G}$.
Let further the function $w = f(z)$ conformally map the domain G onto
some domain D with its boundary also including the straight line part Γ
being the image of the part γ in this mapping: $\Gamma = f(\gamma)$. Then the following
statements are true:

1. *there exists an analytic continuation of the function $f(z)$ from the domain G through the part of the boundary γ into the domain G^* symmetrical to the domain G with respect to γ;*

2. *the analytically continued function conformally maps the domain G^* onto the domain D^* symmetrical to the domain D with respect to Γ;*

3. *if the domains D and D^* do not overlap, then the analytic continuation of the function $f(z)$ conformally maps the domain $G \cup \gamma \cup G^*$ onto the domain $D \cup \Gamma \cup D^*$.*

In the simplest case when γ and Γ are the parts of the real axis, an analytic continuation of the function $f(z)$ onto the domain $G \cup \gamma \cup G^*$ is

$$F(z) = \begin{cases} f(z), & z \in G \cup \gamma, \\ \overline{f(\overline{z})}, & z \in G^*. \end{cases}$$

In the general form of the symmetry principle arcs of circumferences are considered instead of straight line segments γ and (or) Γ and *the symmetry with respect to a circumference* is considered instead of symmetry of the domains G and G^*, D and D^* with respect to straight lines. The symmetry of points of the plane with respect to a circumference is defined as follows. Let l be the circumference of the radius R with the center at some point O. The points M and M^* are called symmetrical with respect to the circumference l if they lie on the same ray starting at the point O and the product of their distances to the point O equals R^2:

$$|OM| \cdot |OM^*| = R^2.$$

The point symmetrical to the center O of the circumference l with respect to the circumference is the point at infinity.

The domains G and G^* are called symmetrical with respect to the circumference l if every point M of one domain is symmetrical to some point M^* of the other domain with respect to the circumference l.

We will assume G to lie to one side of the circumference l (i.e., lies either inside or outside the circumference l). In this case the symmetrical domain G^* lies to the opposite side of the circumference, viz., the domains G and G^* do not intersect.

In the general formulation of the symmetry principle a straight line is considered as the circumference of the extended complex plane crossing through the point at infinity. The symmetry with respect to such a "circumference of the infinite radius" is defined as routine symmetry with respect to the straight line.

Taking into account the concepts introduced the following theorem is formulated.

THEOREM 1.17   *The General Symmetry Principle*
Let the function $f(z)$ be analytic in the domain G with the boundary con-
taining an arc of the circumference γ and continuous in the closed domain
$\overline{G}$. Let further the function $w = f(z)$ map conformally the domain G onto
some domain D with the boundary also containing the arc of the circum-
ference Γ. Let Γ be the image of the arc γ in the mapping: $\Gamma = f(\gamma)$. Then
the following statements are true:

1.   *there exists an analytic continuation of the function $f(z)$ from the*
     domain G through the arc γ into the domain G^* symmetrical to the
     domain G with respect to the arc γ;

2.   *the analytically continued function $f(z)$ conformally maps the domain*
     G^* onto the domain D^* symmetrical to the domain D with respect to
     the arc Γ;

3.   *if the domains D and D^* do not overlap, then the analytic continu-*
     ation of the function $f(z)$ conformally maps the domain $G \cup \gamma \cup G^*$
     onto the domain $D \cup \Gamma \cup D^*$.

1.7   Conformal Mappings Realized by the Basic Elementary Functions

1.7.1   Linear function

A linear function is

$$w = az + b, \tag{1.7.1}$$

where a and b are arbitrary complex constants. When $a \neq 0$ the function
does not degenerate into a constant and has the single-valued inverse one:

$$z = (w - b)/a.$$

Therefore the linear function (1.7.1) is univalent on the complex plane $\mathcal{C}$
and maps $\mathcal{C}$ onto $\mathcal{C}$ conformally. The reverse statement is also true: if the
function $w = f(z)$ maps the open complex plane $\mathcal{C}$ onto $\mathcal{C}$ conformally, it is
a linear one. Indeed, it is necessary for such a mapping that the function
$f(z)$ is entire and has the first order pole at the point at infinity. In this
case it acquires the form

$$f(z) = az + \varphi(z),$$

where $\varphi(z)$ is an entire function without any singularity at the point at
infinity. According to the Liouville's theorem 1.4, such a function can be
nothing but a constant.

The general linear function (1.7.1) can be represented as a sequence of the three simplest transformations:

$$\eta = kz, \quad \xi = e^{i\alpha}, \quad w = \xi + b,$$

where $k = |a|$, $\alpha = \arg a$. The first of these transformations represents a stretching of an arbitrary figure by the factor k, the second — a rotation through the angle α and the third — a parallel translation through the displacement b. It is important that the values k, α and b do not depend on z, so the linear mappings (1.7.1) (and only them) possess the property of form preservation for arbitrary figures, not only infinitesimal ones.

The properties of linear mappings are as follows.

1. The property of straight line preservation. Any straight line of the z-plane is mapped by the function (1.7.1) onto the straight line of the w-plane. The reverse statement is also true: if an analytic function maps any straight line onto a straight line then it is linear.

2. Linear transformation is uniquely determined by the correspondence of two pairs of points. The linear function mapping the points z_1, z_2 into the points w_1 and w_2 correspondingly is determined from the equation

$$\frac{w - w_1}{w_2 - w_1} = \frac{z - z_1}{z_2 - z_1}.$$

The fraction

$$\frac{z - z_1}{z_2 - z_1} = T(z, z_1, z_2) \tag{1.7.2}$$

is called the *ratio of the three points* z, z_1, z_2. It is an invariant of the linear transformation (1.7.1). The three given points z_1, z_2, z_3 can be transformed into the three given points w_1, w_2, w_3 by the linear transformation only in the case when both of these sets of numbers have the same ratios. Geometrically it means that the triangle with the vertexes w_1, w_2, w_3 should be similar to the triangle with the vertexes z_1, z_2, z_3 and oriented identically.

3. Invariant sets of linear transformation. The points z and w being considered as those of the same plane, the question about stationary points and invariant sets of the transformation $w = f(z)$ naturally arises. Stationary points of the transformations are the points satisfying the condition $f(z) = z$. For the linear transformation (1.7.1) $z^* = \infty$ is one of the stationary points for which $w = \infty$. Another stationary point is defined from the equation

$$az + b = z,$$

that yields

$$z^{**} = \frac{b}{1-a}. \tag{1.7.3}$$

As $b = 0$, $a = 1$ the mapping (1.7.1) is an identical transformation $w = z$ for which all the points of the complex plane are stationary. For all the other cases the formula (1.7.3) determines the second stationary point of the linear mapping (as $a = 1$, $b \neq 0$ it coincides with the point $z = \infty$).

When $a = 1$ the transformation (1.7.1) takes the form

$$w = z + b, \tag{1.7.4}$$

representing the parallel translation of points through the displacement b. Any straight line parallel to the vector b is mapped onto itself under this transformation. Such a straight line is an invariant set of the transformation (1.7.4).

As $a \neq 1$ the transformation (1.7.1) has the finite stationary point (1.7.3). If the origin is transferred into the point z^{**}, i.e., new variables are introduced as

$$W = w - z^{**}, \quad Z = z - z^{**},$$

then the transformation (1.7.1) takes the form

$$W = aZ. \tag{1.7.5}$$

Such a transformation can be classified according to a-value as follows:

(a) $a = k$ $(k > 0, k \neq 1)$. In this case the transformation (1.7.5) is a similitude mapping with the center at the point $Z = 0$ and the stretch coefficient k. The straight lines coming through the origin are invariant with respect to such transformations.

(b) $a = e^{i\alpha}$ $(\alpha \neq 0)$. In this case the transformation (1.7.5) is a rotation about the origin. The circumferences with the center at the origin are invariant with respect to this transformation.

(c) $a = k e^{i\alpha}$ $(k > 0, k \neq 1, \alpha \neq 0)$. In this case the transformation (1.7.5) is a combination of a rotation through the angle α and a stretch by the factor of k. The logarithmic spirals* $\rho = C e^{\beta\varphi}$ are invariant with respect to this transformation, where ρ, φ are the polar coordinates of the point z; $\beta = \ln k / \alpha$.

---

*A *logarithmic spiral* is a plane curve with the polar equation $\rho = \exp(\beta\varphi)$, where β is a real constant, $\beta \neq 0$. The logarithmic spiral intersects the rays starting at the origin with the constant angle.

1.7.2 Linear-fractional function

A linear-fractional function is the function

$$w = \frac{az + b}{cz + d}, \tag{1.7.6}$$

where a, b, c, d are arbitrary complex constants. The matrix of coefficients

$$A = \begin{pmatrix} a & b \\ c & d \end{pmatrix}$$

is called the matrix of the linear-fractional function (1.7.6). If it is not singular, i.e., if

$$\Delta = \det A = ad - bc \neq 0,$$

then the function (1.7.6) does not degenerate into a constant. Further below only nondegenerate functions (1.7.6) are considered. As $c = 0$ the function turns into the linear function (1.7.1). If $c \neq 0$ the function (1.7.6) has the first order pole at the point $z_0 = -d/c$, i.e., $w(z_0) = \infty$; the value of the function at the point at infinity is defined via the continuity: $w(\infty) = a/c$.

The inverse mapping of the w-plane onto the z-plane is carried out by the linear-fractional function

$$z = \frac{dw - b}{-cw + a}$$

with the matrix

$$\begin{pmatrix} d & -b \\ -c & a \end{pmatrix}$$

proportional to the matrix A^{-1}. Therefore the function (1.7.6) is analytic and univalent on the extended complex plane $\overline{C}$; it conformally maps $\overline{C}$ onto $\overline{C}$. The reverse is also true: if a function $w = f(z)$ maps conformally the extended complex plane $\overline{C}$ onto $\overline{C}$, it is linear-fractional. Indeed, the function $f(z)$ should be analytic everywhere in $\overline{C}$ except for the only point z_0 in which the function must have the pole of the first order. If $z_0 = \infty$ the function $f(z)$ is linear, being a particular case of the linear-fractional one. If $z_0 \neq \infty$ the function $f(z)$ takes the form:

$$f(z) = \frac{\alpha}{z - z_0} + \varphi(z), \tag{1.7.7}$$

where $\varphi(z)$ is an entire function without singularities in $\overline{C}$. According to the Liouville's theorem such a function can be nothing but a constant. As $\varphi(z) = C$ the function (1.7.7) represents the linear-fractional one.

If the linear-fractional function (1.7.6) is not linear, i.e., if $c \neq 0$, then it can be represented as

$$w = \frac{ad - bc}{c(cz + d)} + \frac{a}{c} = \frac{A}{z + C} + B,$$

where

$$A = \frac{ad - bc}{c^2}, \quad B = \frac{a}{c}, \quad C = \frac{d}{c}.$$

This transformation represents the sequence of three transformations

$$\xi = z + C, \quad Z = \frac{1}{\xi}, \quad w = AZ + B,$$

the first and the last being linear ones. The transformation $w = 1/z$ is the simplest linear-fractional transformation different from the linear one.

Let us enumerate the basic properties of linear-fractional mappings.

1. The group property. If the substitution of the independent z-variable is made in the function (1.7.6), a linear-fractional function will be obtained again. Indeed, if

$$w = \frac{az + b}{cz + d}$$

is the linear-fractional function with the matrix

$$A = \begin{pmatrix} a & b \\ c & d \end{pmatrix}$$

and

$$z = \frac{\alpha \xi + \beta}{\gamma \xi + \delta}$$

is the linear-fractional function with the matrix

$$B = \begin{pmatrix} \alpha & \beta \\ \gamma & \delta \end{pmatrix},$$

then

$$w = \frac{(a\alpha + b\gamma)\xi + (a\beta + b\delta)}{(c\alpha + d\gamma)\xi + (c\beta + d\delta)}$$

representing a linear-fractional function with a non-singular matrix

$$AB = \begin{pmatrix} a & b \\ c & d \end{pmatrix} \cdot \begin{pmatrix} \alpha & \beta \\ \gamma & \delta \end{pmatrix}.$$

Therefore the set of nondegenerate linear-fractional transformations forms the group where the composition of transformations serves as the group operation.

2. The circular property. A circumference of the z-plane is mapped onto a circumference or a straight line on the w-plane under the linear-fractional transformation (1.7.6), a straight line is also mapped onto a circumference or a straight line. Only those circumferences and straight lines of the z-plane are mapped onto straight lines that pass through the point $z_0 = -d/c$ being the preimage of the point at infinity.

In the theory of conformal mappings a straight line is considered as a particular case of the circumference coming through the point at infinity. The circular property of a linear-fractional mapping can be formulated shorter with this terminology: under the transformation (1.7.6) any circumference is mapped onto the circumference. The reverse statement is also valid: if the analytic function $w = f(z)$ maps an arbitrary circumference onto a circumference, then the function is a linear-fractional one.

3. The property of the preservation of symmetry with respect to a circumference. The concept of the symmetry of points with respect to a circumference introduced in the Subsection 1.6.6 geometrically can be reformulated for the points in the complex plane in such a way that the points z and z^* are symmetrical with respect to the circumference γ with the radius R and the center at the point z_0 if

$$\arg(z - z_0) = \arg(z^* - z_0), \quad |z - z_0| \cdot |z^* - z_0| = R^2.$$

It follows from this definition:

$$z^* - z_0 = \frac{R^2}{\overline{z} - \overline{z_0}}. \tag{1.7.8}$$

The linear-fractional transformation preserves the symmetry of points with respect to a circumference. It means that if the points z and z^* of the complex z-plane are symmetrical with respect to some circumference γ, then the linear-fractional function (1.7.6) maps them into points $w = f(z)$, $w^* = f(z^*)$ symmetrical with respect to the image of the circumference γ.

4. The connection of linear-fractional mappings with the inversion. The geometrical transformation when for each point of the z-plane the point z^* is assigned into correspondence which is symmetrical to it with respect to the circumference γ is called an *inversion* with respect to the circumference. In accordance to the formula (1.7.8) the inversion transformation is carried out by the non-analytic function

$$z^* = z_0 + \frac{R^2}{\overline{z} - \overline{z_0}},$$

where R is the radius of the circumference and z_0 is its center. In particular, the inversion with respect to the unit circumference $|z| = 1$ is closely connected with the simplest linear-fractional function:

$$z^* = \frac{1}{\overline{z}}.$$

The inversion transformation is a conformal mapping of the second class carried out by the function which is a complex conjugate to the linear-fractional one. However, further we call the simplest linear-fractional mapping $w = 1/z$ the *inversion* for brevity.

5. A linear-fractional mapping is uniquely defined by correspondence of three pairs of points. Let the three different points z_1, z_2, z_3 be given in the z-plane and their images w_1, w_2, w_3 be given in the w-plane. The planes z and w can be mapped onto the auxiliary t-plane with the correspondence of the points as:

z	z_1	z_2	z_3
w	w_1	w_2	w_3
t	0	1	∞

The mapping of the z-plane onto the t-plane is carried out by the function

$$t = A \, \frac{z - z_1}{z - z_3},$$

where

$$A = 1 : \frac{z_2 - z_1}{z_2 - z_3}.$$

Analogously the mapping of the w-plane onto the t-plane is realized by the function

$$t = \frac{w - w_1}{w - w_3} : \frac{w_2 - w_1}{w_2 - w_3}.$$

Eliminating the auxiliary variable t the sought-for function $w(z)$ is obtained in implicit form

$$\frac{w - w_1}{w - w_3} : \frac{w_2 - w_1}{w_2 - w_3} = \frac{z - z_1}{z - z_3} : \frac{z_2 - z_1}{z_2 - z_3}. \qquad (1.7.9)$$

The expression

$$D(z, z_1, z_2, z_3) = \frac{z - z_1}{z - z_3} : \frac{z_2 - z_1}{z_2 - z_3} \qquad (1.7.10)$$

is called an *anharmonic (or cross-) ratio* of the four points. The equality (1.7.9) means that the anharmonic ratio is invariant for arbitrary four points under the linear-fractional mapping. The four points z_1, z_2, z_3, z_4 can be mapped by the linear-fractional function into the four points w_1, w_2, w_3, w_4 when and only when these sets of four numbers have the same anharmonic ratio.

6. Stationary points and invariant sets of linear-fractional mappings.

Let us consider the points z and w as those of the same plane and define stationary points of the mapping (1.7.6) when $c \neq 0$ (the case $c = 0$ was considered above in the Subsection 1.7.1). The equation of stationary points $z = (az + b)/(cz + d)$ in this case can be reduced to the quadratic one with the roots

$$z_{1,2} = \frac{a - d + \sqrt{(a-d)^2 + 4bc}}{2c}. \qquad (1.7.11)$$

If $(a - d)^2 + 4bc \neq 0$ then the transformation (1.7.6) has two different stationary points z_1 and z_2 defined by (1.7.11). In this case the linear-fractional substitution of variables

$$Z = \frac{z - z_1}{z - z_2}, \quad W = \frac{w - z_1}{w - z_2} \qquad (1.7.12)$$

maps the point z_1 into the origin and the point z_2 into infinity. Because of the group property the function $W = W(Z)$ is linear-fractional. Its stationary points are $Z = 0$ and $Z = \infty$. Such a function is only the linear one (1.7.5):

$$W = AZ,$$

considered above in the Subsection 1.7.1. Therefore the canonical form of the linear-fractional transformation with two stationary points is:

$$\frac{w - z_1}{w - z_2} = A \frac{z - z_1}{z - z_2}.$$

The coefficient A can be expressed through the coefficients of the linear-fractional transformation (1.7.6) as follows:

$$A = \frac{a + d + \sqrt{(a-d)^2 + 4bc}}{a + d - \sqrt{(a-d)^2 + 4bc}}.$$

The preimage of the polar net of lines under the mapping (1.7.12) is the so-called *bipolar net*. Is consists of two reciprocally-orthogonal families of circumferences. The circumferences of the z-plane passing through the points z_1 and z_2 are the preimages of the straight lines passing through the point $Z = 0$. These circumferences are called *Steiner's* ones, their arcs connecting the points z_1 and z_2 are defined by the equations

$$\arg \frac{z - z_1}{z - z_2} = \text{const}.$$

The preimages of the circumferences $|Z| = C$ are the circumferences orthogonal to the Steiner's ones. These lines are called the *Apollonian circumferences*; they are defined by the equations

$$\left| \frac{z - z_1}{z - z_2} \right| = \text{const}.$$

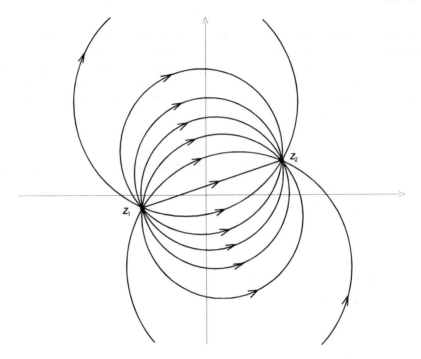

FIGURE 1.7

The linear-fractional transformations with two stationary points are classified similarly to the linear transformations (1.7.5) according to the value of A-coefficient:

(a) When $A = k$ $(k > 0,\ k \neq 1)$ the linear-fractional transformation is called the hyperbolic one. In this case the straight lines passing through the origin are invariant for the transformation (1.7.5). Their preimages, i.e., Steiner's circumferences passing through the two stationary points z_1 and z_2 are invariant for the linear-fractional transformation (1.7.6). The family of such circumferences is shown in the Fig. 1.7

(b) When $A = e^{i\alpha}$ $(\alpha \neq 0)$ the linear-fractional transformation is called the elliptic one. Concentric circumferences with the center at the origin are invariant for the linear transformation (1.7.5). Their preimages in the z-plane are Apollonian circumferences with the limiting points z_1 and z_2. Under this transformation each of the Apollonian circumferences is mapped onto itself. A family of such circumferences is depicted in Fig. 1.8.

(c) When $A = k\,e^{i\alpha}$ $(k > 0,\ k \neq 1,\ \alpha \neq 0)$ the linear-fractional transformation is called the loxodromic one. Logarithmic spirals are in-

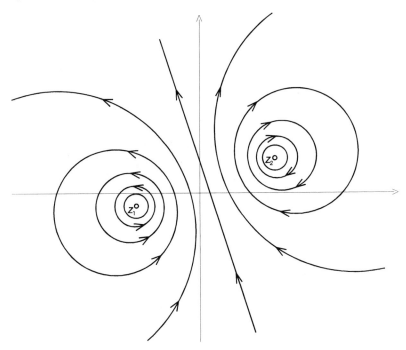

FIGURE 1.8

variant with respect to the linear transformation (1.7.5). For the linear-fractional transformation (1.7.6) invariant curves are preimages of logarithmic spirals called loxodromes.[†] A family of loxodromes is depicted in Fig. 1.9.

Finally in the case $(a - d)^2 + 4bc = 0$ when roots of the quadratic equation (1.7.11) coincide, the linear-fractional transformation is called the parabolic one. In this case the linear-fractional change of variables

$$Z = \frac{1}{z - z_1}, \quad W = \frac{1}{w - z_1} \tag{1.7.13}$$

maps the point z_1 into infinity. The function $W = W(Z)$ represents the linear-fractional transformation with a two-fold stationary point at infinity. The linear transformation

$$W = Z + h \tag{1.7.14}$$

is the only transformation representing a parallel transition along a vector h. Therefore the canonical form of the linear-fractional transformation with

---

[†]The term "loxodrome" came from the cartography. It is the name of curves on the sphere crossing meridians with the constant angle (isogonally). Loxodromes in a plane are the curves crossing the Steiner's circumferences with the constant angle.

FIGURE 1.9

one finite two-fold stationary point z_1 is:

$$\frac{1}{w - z_1} = \frac{1}{z - z_1} + h.$$

The value h is expressed through coefficients of the linear-fractional transformation (1.7.6) as follows:

$$h = \frac{2c}{a + d}.$$

Straight lines parallel to the vector h are invariant with respect to the transformation (1.7.14). Under the transformation (1.7.13) preimages of straight lines are circumferences passing through the point z_1 and touching at this point the straight line directed along the vector

$$H = \frac{1}{h} = \frac{a + d}{2c}.$$

Each of these circumferences is mapped onto itself under the parabolic transformation. A family of these circumferences is depicted in Fig. 1.10.

7. A relief map and an isothermic net of the linear-fractional function. A relief map of the function (1.7.6) is built as the image of the

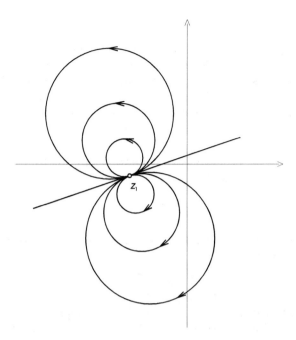

FIGURE 1.10

polar net under the transformation carried out by the inverse function (see Section 1.3):

$$z = \frac{dw - b}{-cw + a}. \tag{1.7.15}$$

The point $w = 0$ is mapped by the function (1.7.15) into the point $z_0 = -b/a$ being the zero of the function (1.7.6) (if $a \neq 0$). The preimage of the point $w = \infty$ is the point $z_\infty = -d/c$ being the pole of the function (1.7.6). If $ac \neq 0$ the function (1.7.15) maps the polar net of lines onto the bipolar one built on the points z_0 and z_∞. Therefore if $ac \neq 0$ the relief map of the linear-fractional function (1.7.6) represents the bipolar net of lines. For the Apollonian circumferences we have $|w| = $ const, for the Steiner's circumferences (more precisely on their arcs connecting the points z_0 and z_∞) we have $\arg w = $ const. At last if $ac = 0$ the relief map of the linear-fractional function represents a polar net of lines with the center at the point z_0 (if $a \neq 0$) or at the point z_∞ (if $c \neq 0$).

The isothermic net of the function (1.7.6) represents an image of a Cartesian net under the inverse mapping (1.7.15). If $c \neq 0$ (i.e., in the case when the function (1.7.6) is not linear), a family of parallel straight lines is mapped by the function (1.7.15) onto a family of circumferences passing through the point $z_\infty = -d/c$ and having a common tangent in it.   The Cartesian net of lines is mapped by the function (1.7.15) onto two reciprocally-orthogonal families of circumferences passing through the point z_∞. Such a net of circumferences we name the *unipolar net*. The straight lines $v = $ const are mapped by the function (1.7.15) onto the circumferences with the common tangent having an inclination angle equal $\arg(\Delta/c^2)$. The straight lines $u = $ const are mapped onto the circumferences with the common tangent having an inclination angle equal $\arg(\Delta/c^2) + \pi/2$.

In conclusion let us consider four linear-fractional mappings frequently met with in applications.

Example 1.5

The mapping of the real axis z onto the real axis w is carried out by the linear-fractional function

$$w = \frac{az + b}{cz + d}$$

with the real coefficients a, b, c, d. At that the upper half-plane Im $z > 0$ is mapped onto the upper half-plane Im $w > 0$ if $\Delta = ad - bc > 0$.

Example 1.6

The mapping of the upper half-plane onto the unit circle $|w| < 1$ when the given point z_0 is mapped into the center of the circle is carried out by the function

$$w = e^{i\theta}\, \frac{z - z_0}{z - \overline{z_0}}, \qquad (1.7.16)$$

where $\theta \in \mathcal{R}$. There exists an infinite set of such transformations. If to demand that a linear element at the point z_0 should rotate by the given angle α, the transformation can be defined uniquely:

$$w = i\, e^{i\alpha}\, \frac{z - z_0}{z - \overline{z_0}}.$$

Example 1.7

The mapping of the unit circle $|z| < 1$ onto the unit circle $|w| < 1$ when the point z_0 ($|z_0| < 1$) is mapped into the center of the circle is

$$w = e^{i\theta}\, \frac{z - z_0}{z \cdot \overline{z_0} - 1},$$

where θ is an arbitrary real number. If to demand that a linear element at the point z_0 should rotate by the angle α, the linear-fractional function is defined uniquely:

$$w = e^{i\alpha}\,\frac{z - z_0}{1 - z \cdot \overline{z_0}}. \tag{1.7.17}$$

Example 1.8

Analogously the mapping of the exterior of the unit circle $|z| > 1$ onto the circle $|w| < 1$ can be found, in which the point z_0 ($|z_0| > 1$) is mapped into the center of the circle:

$$w = e^{i\alpha}\,\frac{z - z_0}{z \cdot \overline{z_0} - 1}.$$

_____

Exercises

1. With the help of the program CONFORM create a mapping of the polar net inside the unit circle, carried out by the function being inverse to (1.7.17):

$$z_0 = p\,e^{i\alpha}$$

$$f = \frac{z + z_0}{1 + z\,\overline{z_0}} \tag{1.7.18}$$

for different values $0 < p < 1$. Note the image damaging as p approaches the unit. Establish experimentally up to what values of the variable p it is possible to construct the picture without essential visual distortions. Construct the mapping of the polar net inside the unit circle by the function (1.7.18) with $p > 1$. Take $p = 2$, for example.

2. Create the relief map of the function $w = (z - 1)/(z + 1)$ as the mapping carried out by the inverse function

$$w = \frac{1 + z}{1 - z}. \tag{1.7.19}$$

Take the polar net inside the circle $|z| < 5$, $-\pi < \arg z < \pi$. Take care for a grid point of the net not to hit exactly into the pole of the function (1.7.19), for example, in such a way: $-3.14 < \varphi < 3.141$.

1.7.3 The entire power function and related functions

The function

$$w = z^n, \tag{1.7.20}$$

where n is a natural number, is called the entire power function. If $n = 1$ the function (1.7.20) is a linear one, so only the values $n \geq 2$ are considered here. The function (1.7.20) has the single critical point $z = 0$ and the pole of the n-th order at the point at infinity, therefore the mapping carried out by the function is isogonal at all the points of the plane, except for $z = 0$ and $z = \infty$.

For the function (1.7.20) to be univalent in some domain G it is necessary for this domain not to have any pair of points z_1 and z_2 connected by the relation

$$z_1 = z_2 \, e^{2\pi i/n} \,. \tag{1.7.21}$$

More often an angular domain

$$C < \arg z < C + \frac{2\pi}{n} \tag{1.7.22}$$

is considered as the domain of the univalence of the function (1.7.20). This domain is mapped by the function (1.7.20) conformally onto the domain

$$nC < \arg w < nC + 2\pi \,,$$

which represents the plane w with the cut along a ray. In particular the domain $-\pi/n < \arg z < \pi/n$ is mapped conformally onto the plane w with the cut along the negative real half-axis.

In polar coordinates $z = \rho e^{i\varphi}$ one obtains

$$|w| = \rho^n \,, \quad \arg w = n\varphi \,.$$

Therefore the polar net of lines in the angular domain (1.7.22) is mapped by the function (1.7.20) also into the polar (but deformed) net in the w-plane. The angles at the point $z = 0$ are magnified by a factor of n under the mapping.

Especially consider the case $n = 2$, i.e., the function

$$w = z^2 \,. \tag{1.7.23}$$

The domain of the univalence of the function is the half-plane $C < \arg z < C + \pi$, for example, the right half-plane $\operatorname{Re} z > 0$. A straight line passing through the origin is mapped by this function into the cut along a ray. Any other straight line is mapped into a parabola. A half-plane not containing the origin is mapped conformally onto the exterior of the parabola by the function (1.7.23).

Let us consider the mapping of the Cartesian net carried out by the function $w = z^2$ in more details. In Cartesian coordinates

$$w = u + iv = (x + iy)^2$$

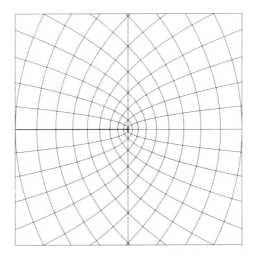

FIGURE 1.11

or

$$\begin{cases} u = x^2 - y^2, \\ v = 2xy. \end{cases}$$

The straight lines $x = C$ are mapped into the curves defined parametrically:

$$u = C^2 - y^2, \quad v = 2Cy.$$

Eliminating the parameter y obtain (if $C \neq 0$) the equation of parabolas

$$u = C^2 - \left(\frac{v}{2C}\right)^2. \tag{1.7.24}$$

Similarly the straight lines $y = C_1$ are mapped (if $C_1 \neq 0$) into a family of parabolas

$$u = \left(\frac{v}{2C_1}\right)^2 - C_1^2, \tag{1.7.25}$$

orthogonal to the family (1.7.24). The Cartesian net is mapped by the function $w = z^2$ into two reciprocally-orthogonal families of confocal parabolas shown in Fig. 1.11.

This net of parabolas can be considered as the isothermic net of the inverse function $w = \sqrt{z}$. Simultaneously it is a net of coordinate lines of the parabolic coordinate system (see Section 1.8 below).

Let us consider the mapping of the polar net with an arbitrary disposed center carried out by the function (1.7.23). For the sake of definiteness let us place the center into the point $z = a$ ($a > 0$) and construct the mapping realized by the function

$$w = (z + a)^2 . \tag{1.7.26}$$

This function is univalent inside the circle $|z| < a$. It maps the circumference $z = a \, e^{i\varphi}$ into the line

$$w = 4a^2 \cos^2(\varphi/2) \, e^{i\varphi} .$$

A plane line with the polar equation $\rho = 2l \cos^2(\varphi/2) = l(1 + \cos\varphi)$ is called a cardioid with the parameter $l = 2a^2$. Thus the circle $|z| < a$ is mapped by the function (1.7.26) onto the plane bounded by the cardioid with the parameter $l = 2a^2$. In the plane $\xi = z + a$ the curve $\xi = a + a \, e^{it}$ $(-\pi < t < \pi)$ is a circumference passing through the origin. Any of such circumferences is mapped by the function $w = \xi^2$ into the cardioid.

Let us consider one more rational function

$$w = \frac{1}{z^2} . \tag{1.7.27}$$

The half-plane $C < \arg z < C + \pi$ is the domain of univalence for this function as well as for that of (1.7.23). The function (1.7.27) can be represented as the composition of the inversion and the quadratic function:

$$\xi = \frac{1}{z}, \quad w = \xi^2 .$$

The inversion maps the Cartesian net of lines into the unipolar net of circumferences passing through the origin. The transformation $w = \xi^2$ maps each circumference of the unipolar net into the cardioid. Therefore the function (1.7.27) maps the Cartesian net into two reciprocally-orthogonal families of cardioids. Such a family of curves is depicted in Fig. 1.12. Otherwise it can be considered as an isothermic net for the inverse function $w = 1/\sqrt{z}$.

The function inverse to $w = z^2$ is the two-valued one. It has two branches

$$w_1 = \sqrt{z}, \quad w_2 = -\sqrt{z} .$$

The points $z = 0$ and $z = \infty$ are its branch points. If a cut is drawn which connects the branch points, each branch is a single-valued analytic function in the singly connected domain obtained. Usually the cut is made along the negative real half-axis, i.e.,

$$-\pi < \arg z < \pi , \quad -\pi/2 < \arg w_1 < \pi/2 .$$

It is this branch that is calculated by computer as the value of the function SQRT(Z). The function $w_1 = \sqrt{z}$ maps the z-plane with the cut onto the right half-plane Re $w > 0$. At that the Cartesian net is mapped into two

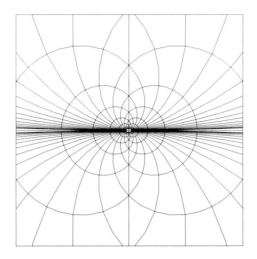

FIGURE 1.12

reciprocally-orthogonal families of equi-sided hyperbolas. Indeed, from the equation $w^2 = z$ we have

$$(u + iv)^2 = x + iy$$

or

$$\begin{cases} u^2 - v^2 = x, \\ 2uv = y. \end{cases}$$

The family of straight lines $x = C$ is mapped into the family of equi-sided hyperbolas $u^2 - v^2 = C$ and the straight lines $y = 2C_1$ are mapped into the family of equi-sided hyperbolas $uv = C_1$.

The image of the Cartesian net under the mapping $w = \sqrt{z}$ is represented in Fig. 1.13. Otherwise the net of lines can be considered as an isothermic one for the inverse function $w = z^2$.

Let us consider the mapping of the polar net with an arbitrary center carried out by the function $w = \sqrt{z}$. Let us place the center into the point $z = a$ $(a > 0)$ and consider the function

$$w = \sqrt{z + a}. \tag{1.7.28}$$

The function (1.7.28) is analytic and univalent inside the circle $|z| < a$. It maps the circumference $z = a\,e^{it}$ into the curve

$$w = \sqrt{2a \cos(t/2)}\, e^{it/4},$$

which can be written as $w = \sqrt{2a \cos(2\varphi)}\, e^{i\varphi}$, where $\varphi = t/4$. The curve with the polar equation $\rho = l\sqrt{\cos(2\varphi)}$ is called *Bernoulli's lemniscata*. Therefore the function (1.7.28) maps conformally the circle $|z| < a$ onto the

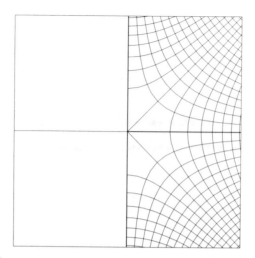

FIGURE 1.13

domain bounded by the loop of Bernoulli's lemniscata with the parameter $l = \sqrt{2a}$. In the plane $\xi = z + a$ the curve $\xi = a + a\,e^{it}$ represents a circumference passing through the origin. Such a circumference is mapped into the loop of Bernoulli's lemniscata by the function $w = \sqrt{\xi}$.

Finally let us construct the isothermic net of the function (1.7.27)

$$w = \frac{1}{z^2}$$

as the image of the Cartesian net under the mapping carried out by the inverse function $w = \pm 1/\sqrt{z}$. The function $w_1 = 1/\sqrt{z}$ can be considered as a composition of two functions:

$$\xi = \frac{1}{z}, \quad w = \sqrt{\xi}.$$

The first of these functions (inversion) maps the Cartesian net into the unipolar net of circumferences passing through the origin. The function $w = \sqrt{\xi}$ maps each circumference into the Bernoulli's lemniscata.

The isothermic net of the function $w = 1/z^2$ representing two reciprocally-orthogonal families of Bernoulli's lemniscatas is depicted in Fig. 1.14.

Exercises

1. With the help of program CONFORM construct a conformal mapping of

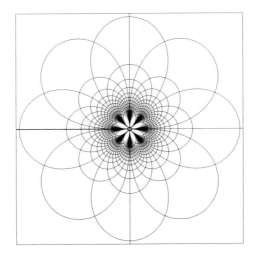

FIGURE 1.14

the circle $|z| < 1$ realized by the function (1.7.26)

$$w = (z+1)^2 .$$

2. Construct at the computer display the isothermic net of the function $w = z^2$, drawing level lines of harmonic conjugate functions $u(x,y)$, $v(x,y)$. Use **Isoline** mode of the program CONFORM. Compare the result with the Fig. 1.13 and explain why distortions appear when the direct construction of level lines is made.

1.7.4 The Zhukovskii function

A rational function

$$w = \frac{z^2 + 1}{2z} = \frac{1}{2} \left(z + \frac{1}{z} \right) \qquad (1.7.29)$$

is called the *Zhukovskii function*. It has first order poles at the points $z = 0$ and $z = \infty$. The derivative of the function (1.7.29)

$$w'(z) = \frac{1}{2} \left(1 - \frac{1}{z^2} \right)$$

vanishes at the points $z = \pm 1$. The points $z = \pm 1$ are the critical ones, the mapping (1.7.29) is isogonal at all the other points of the extended plane $\overline{\mathcal{C}}$.

The function (1.7.29) is univalent in any domain not including points z_1 and z_2 connected by the relation $z_1 z_2 = 1$. Indeed, at points z_1 and $z_2 = 1/z_1$ the function (1.7.28) has the same values. The opposite is also

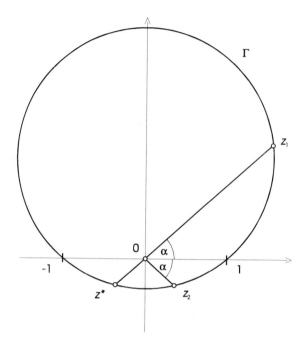

FIGURE 1.15

true: if $w(z_1) = w(z_2)$ and $z_1 \neq z_2$ then $z_1 z_2 = 1$.[‡]

The point $z_2 = 1/z_1$ can be geometrically constructed in such a way as is shown in Fig. 1.15. Here Γ designates a circumference passing through the points -1, 1 and z_1 (z_1 is assumed not to lie on the real axis), z^* is the end of the chord $z_1 z^*$ passing through the origin, and z_2 is the point being symmetrical to z^* with respect to the imaginary axis, $\alpha = \arg z$. Two chords intersect in the origin: $[z_1, z^*]$ and $[-1, 1]$. The known theorem concerning the intersecting chords yields $|z_1| |z^*| = 1$. For the symmetrical point z_2 we have

$$|z_2| = \frac{1}{|z_1|}, \quad \arg z_2 = -\alpha = -\arg z_1,$$

it means that $z_2 = 1/z_1$. It can be seen from the Fig. 1.15 that any circumference Γ passing through the points 1 and -1 separates two domains of univalence of the Zhukovskii function. In particular, univalence domains are: the interior of the unit circle $|z| < 1$, the exterior of the circle $|z| > 1$, the upper half-plane $\text{Im } z > 0$, and the lower half-plane $\text{Im } z < 0$.

Let Γ be the circumference passing through the points 1 and -1 and intersecting the real axis at an angle β (Fig. 1.16). Let us consider what is

---

[‡]Indeed, from the equality $z_1 + 1/z_1 = z_2 + 1/z_2$ there follows $(z_1 - z_2)(1 - 1/(z_1 z_2)) = 0$ and hence $z_1 z_2 = 1$.

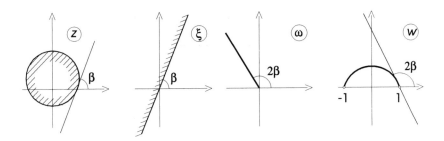

FIGURE 1.16

the domain the function (1.7.29) maps the exterior of the circumference Γ onto. Rewriting the Zhukovskii function in the form

$$w = \frac{1 + \left(\frac{z-1}{z+1}\right)^2}{1 - \left(\frac{z-1}{z+1}\right)^2}$$

let us represent it as composition of three simple functions

$$\xi = \frac{z-1}{z+1}, \quad \omega = \xi^2, \quad w = \frac{1+\omega}{1-\omega}.$$

The exterior of the circumference Γ is mapped onto the half-plane $\beta - \pi <$ $\arg \xi < \beta$ under the linear-fractional transformation $\xi = (z - 1)/(z + 1)$. The quadratic transformation $\omega = \xi^2$ maps this half-plane onto the plane with the cut along the ray $\arg \omega = 2\beta$. At last, under the linear-fractional transformation $w = (1 + \omega)/(1 - \omega)$ the ray is mapped onto the arc of a circumference with the ends at the points $w = \pm 1$ and the angle 2β with the real axis. As the result the Zhukovskii function maps conformally the exterior of Γ onto the w-plane with the cut along the mentioned arc. The interior of the circumference Γ is mapped onto just the same domain.

In particular, the circumference Γ represents the unit circumference $|z| = 1$ as $\beta = \pi/2$. The Zhukovskii function maps conformally the exterior of the unit circle onto the w-plane with the cut along the segment of the real axis $[-1, 1]$. Under that the polar net in the exterior of the circle is mapped onto the net of confocal ellipses and hyperbolas with foci at the points $w = \pm 1$. Such a net of lines is depicted in Fig. 1.17.

The Zhukovskii function is univalent in the half-plane $\text{Im } z > 0$. It maps

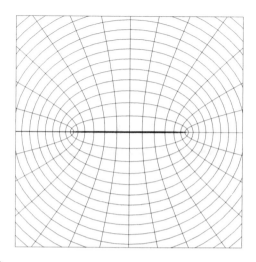

FIGURE 1.17

the half-plane onto the w-plane with two cuts along the rays of the real axis $(-\infty, -1)$ and $(1, \infty)$. The mapping of the Cartesian net in the half-plane Im $z > 0$ is presented in Fig. 1.18.

Each circumference of the z-plane in the univalence domain is mapped by the function (1.7.29) into the closed Jordan curve called the *Zhukovskii profile*. If the circumference passes through one of the points 1 or -1, then the Zhukovskii profile has a cusp (the returning point), in the opposite case it represents a smooth Jordan curve.

A family of Zhukovskii profiles is depicted in Fig. 1.19. Zhukovskii profiles are widely used in aerodynamics for aerofoil profiles approximation. More about this is in Subsection 2.18.3.

Function inverse to the Zhukovskii function is two-valued. It has two branches:

$$w_1 = z + \sqrt{z^2 - 1} \quad \text{and} \quad w_2 = z - \sqrt{z^2 - 1}. \qquad (1.7.30)$$

The points $z = \pm 1$ are branch points. If a cut is made which connects branch points, analytic single valued branches of the inverse function can be separated in the obtained domain.

The mapping of a polar net carried out by the inverse function is represented in Fig. 1.20. This figure is the relief map of the Zhukovskii function.

The isothermic net of the Zhukovskii function is depicted in Fig. 1.21. It represents a mapping of a Cartesian net carried out by the inverse function (1.7.30).

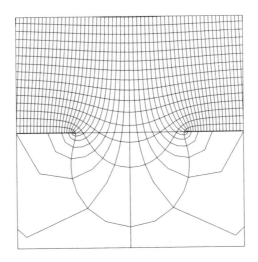

FIGURE 1.18

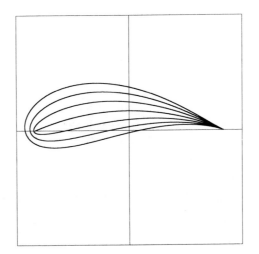

FIGURE 1.19

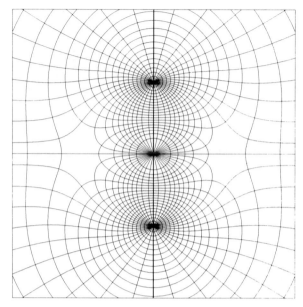

FIGURE 1.20

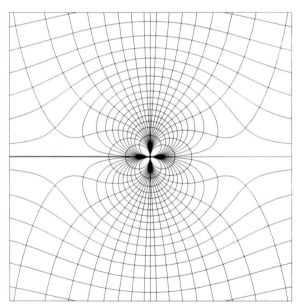

FIGURE 1.21

Exercises

1. With the help of the program CONFORM create the mapping of the polar net in the exterior of the circle $|z| > 1$ carried out by the function

$$s = \left(1 + \frac{p_1 + p_2}{2}\right) z + \frac{p_2 - p_1}{2},$$

$$w = \frac{1}{2}\left(s + \frac{1}{s}\right)$$

   with $p_1 = 0.35$ and various values of the parameter p_2 $(0 < p_2 < 0.1)$. Make sure that the circumference $|z| = 1$ is mapped into the closed curve, namely the symmetrical Zhukovskii profile (in other words the *Zhukovskii rudder*). If $p_2 = 0$ the curve has a sharp edge — the returning point. If $p_2 > 0$ Zhukovskii profile has the rounded off edge instead of the sharp one.

2. Create the mapping of the polar net in the exterior of the unit circle $|z| > 1$ carried out by the function

$$s = \frac{z + i \cos p}{\sin p},$$

$$w = \frac{1}{2}\left(s + \frac{1}{s}\right)$$

   for various values of the parameter p, $\pi/4 < p < \pi/2$. Make sure that if $p = \pi/4$ the w-plane with the cut along an arc of semi-circumference and if $p = \pi/2$ the plane with the cut along the segment $[-1, 1]$ is the image.

1.7.5 The exponential function

The exponential function of the complex variable $z = x + iy$ is defined as

$$w = e^z = e^x\, e^{iy} = e^x\, \cos y + i\, e^x\, \sin y. \tag{1.7.31}$$

This function satisfies the Cauchy-Riemann conditions (1.2.2). It is equal to e^x on the real axis, thus it represents an analytic continuation of the function e^x to the whole complex plane $\mathcal{C}$.

The function e^z is an entire transcendental function (i.e., it has an essential singularity at the point at infinity). Its modulus and argument are defined as

$$|w| = e^x, \quad \arg w = y, \tag{1.7.32}$$

therefore, the relief map of the function e^z represents the Cartesian net in the z-plane. The isothermic net of the function e^z will be described in the next subsection.

The derivative of the function (1.7.31) $w' = e^z$ does not vanish, therefore, the mapping $w = e^z$ is locally conformal at any point of the complex plane $\mathcal{C}$.

It follows from the definition of the function (1.7.31) that it is the periodic one with a pure imaginary period $2\pi i$, i.e., $e^z = e^{z+2\pi i n}$ for any $n \in \mathcal{Z}$.

The reverse is also true: if $e^{z_1} = e^{z_2}$ then $z_2 = z_1 + 2\pi i n$, $n \in \mathcal{Z}$. Therefore, the function e^z is univalent in any domain not containing the points z_1 and z_2 connected by relation $z_2 = z_1 + 2\pi i$. A curvilinear strip

$$f(x) < y < f(x) + 2\pi \qquad (1.7.33)$$

can be such a domain, where $f(x)$ is an arbitrary continuous function defined on the whole numerical axis $\mathcal{R}$. More often the horizontal strip $C < \operatorname{Im} z < C + 2\pi$ is considered as the domain of univalence. The image of this strip in the mapping (1.7.31) is the whole complex plane w with the cut along the ray $\arg w = C$. In particular, the strip $-\pi < \operatorname{Im} z < \pi$ is conformally mapped by the function (1.7.31) onto the w-plane with the cut along the negative real half-axis and the strip $0 < \operatorname{Im} z < \pi$ is conformally mapped onto the upper half-plane $\operatorname{Im} w > 0$. The function $w = e^z$ is always used when it is necessary to map conformally the strip $0 < \operatorname{Im} z < \pi$ onto a half-plane. At that a Cartesian net of lines in the strip is mapped onto the polar net in the half-plane, namely horizontal straight lines $\operatorname{Im} z = C_1$ are mapped into the rays $\arg w = C_1$ and segments of vertical straight lines $\operatorname{Re} z = C$ are mapped into semi-circumferences $|w| = e^C$, $\operatorname{Im} w > 0$.

Let us consider into what line the function (1.7.31) maps the inclined straight line $y = ax + b$. Rewriting the function w in the exponential form

$$w = R e^{i\Phi}$$

we have according to (1.7.32) that $R = e^x$, $\Phi = ax + b$. Eliminating the parameter x we obtain

$$R = \exp \frac{\Phi - b}{a}.$$

Thus any inclined straight line of the z-plane is mapped by the function (1.7.31) into the logarithmic spiral. If the straight line makes the angle α with the abscissa axis ($\alpha = \arctan a$), the logarithmic spiral intersects its radius-vectors with the same angle α because of conformity of the mapping. A family of parallel inclined straight lines corresponding to various values of the parameter p is mapped into a family of congruent logarithmic spirals

$$R = C \exp\left(\frac{\Phi}{a}\right), \quad \text{where } C = \exp\left(-\frac{b}{a}\right). \qquad (1.7.34)$$

According to the condition (1.7.33) the function $w = e^z$ is univalent in the inclined strip

$$ax - \pi < y < ax + \pi.$$

Otherwise these conditions can be written as $|y - ax| < \pi$ or $-\pi < \operatorname{Im}(z(1 - ia)) < \pi$. Designating $\xi = z(1 - ia)$ we obtain that the function

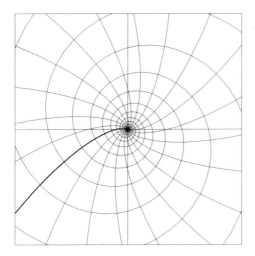

FIGURE 1.22

$w = \exp(\xi/(1 - ia))$ maps the horizontal strip $-\pi < \operatorname{Im} \xi < \pi$ onto the w-plane with a cut along the logarithmic spiral

$$R = \exp\left(\frac{\Phi - \pi}{a}\right).$$

The Cartesian net inside the horizontal strip $-\pi < \operatorname{Im} \xi < \pi$ is mapped by the function $w = \exp(\xi/(1 - ia))$ into two reciprocally-orthogonal families of logarithmic spirals, namely horizontal straight lines are mapped into the family of infinite spirals (1.7.34) and segments of vertical straight lines $\operatorname{Re} \xi = C$ are mapped into arcs of logarithmic spirals orthogonal to the family (1.7.34).

The mapping of the Cartesian net carried out by the function

$$w = \exp\left(\frac{z}{1 - ia}\right)$$

with $a = 0.25$ is presented in Fig. 1.22

Exercises

1. Find the maximum value of R the function $w = e^z$ to be univalent inside the circle $|z| < R$. With the help of the program CONFORM create a mapping of the polar net in the circle $|z| < R$. Interpret the obtained net of lines as a relief map of the inverse function $w = \ln z$ in some domain.
2. With the help of the program CONFORM (**Isoline** mode) create an

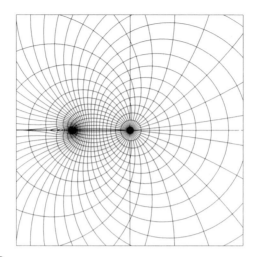

FIGURE 1.23

isothermic net of the function $w = e^z$. Compare the result with the Fig. 1.24.

1.7.6   Logarithmic function

The function inverse to the infinitely-valent function e^z is infinitely-valued function

$$\operatorname{Ln} z = \ln |z| + i \operatorname{Arg} z,$$

where $\operatorname{Arg} z = \arg z + 2\pi n$, $n \in \mathcal{Z}$. Its branch points are $z = 0$ and $z = \infty$. A single-valued continuous branch of the analytic function can be separated by making a cut connecting the branch points. Usually the cut is made along the negative real half-axis and the principal value of the logarithmic function is designated as $\ln z$, hence

$$\ln z = \ln |z| + i \arg z.$$

It is the value the computer evaluates as the function $\operatorname{LOG}(Z)$ value. The function $w = \ln z$ is analytic and univalent in the plane $\mathcal{C}$ with the cut along the negative real half-axis. This function conformally maps the plane with the cut onto the strip $-\pi < \operatorname{Im} w < \pi$ and the upper half-plane $\operatorname{Im} z > 0$ onto the strip $0 < \operatorname{Im} w < \pi$. At that the polar net of lines is mapped into the Cartesian net in the strip.

The isothermic net of the function $w = \ln z$ is the polar net. The relief map of the function $w = \ln(z + 1)$ is presented in Fig. 1.23

Let us consider the mapping of the Cartesian net carried out by the function $w = \ln z$. The vertical straight lines $x = C$ have the polar equation

$\rho = C / \cos \varphi$. They are mapped into lines with the parametric equation

$$u = \ln \left(\frac{C}{\cos \varphi} \right) , \quad v = \varphi .$$

Excluding the parameter φ an explicit equation of lines is obtained:

$$u = - \ln(\cos v) + \ln C . \tag{1.7.35}$$

The line defined by the equation

$$y = - \ln \cos x , \quad -\frac{\pi}{2} < x < \frac{\pi}{2} , \tag{1.7.36}$$

is called a catenary of a constant resistance. This shape is acquired by flexible heavy fiber with varying cross-section under gravity force. The dependence of the cross-section area on the length should be such that the tension force density would be the same in all sections. In this fiber the probability of breakage (or "the resistance to breakage") at all its points is constant;[§] this explains the name of the curve (1.7.36).

Thus, the straight lines $x = C$ are mapped by the function $w = \ln z$ into the catenaries of the constant resistance (1.7.35). Any straight line not passing through the origin is mapped into the similar but differently disposed catenary.

The mapping of the Cartesian net carried out by the function $w = \ln z$ is depicted in Fig. 1.24. This Figure simultaneously presents the isothermic net of the inverse function $w = e^z$.

In conclusion let us consider into which net of lines the function $w = \ln z$ maps the unipolar net of circumferences. For the purpose it can be represented as the composition of the two functions

$$\xi = \frac{1}{z} , \quad w = - \ln \xi .$$

The first function (inversion) maps the Cartesian net into the unipolar one. The second function $w = - \ln \xi$ maps the unipolar net into that of catenaries of constant resistance depicted in Fig. 1.24.

Therefore, the function $w = \ln z$ maps the unipolar net into the net of catenaries (Fig. 1.24) turned by 180^0.

1.7.7 The general power function

For non-integer values of the index α the function z^α is defined as $\exp(\alpha \operatorname{Ln} z)$. Together with $\operatorname{Ln} z$ it is multi-valued function with branch

---

[§] Indeed, let the shape of the fiber be determined by the equation $y = - \ln \cos x$, then the length of an arc element is equal to $ds = dx / \cos x$. Let the cross-section area change along the fiber as $S(x) = S / \cos x$. Then such a fiber is in the equilibrium state under tension force $T(x) = T / \cos x$ and gravity proportional to $S(x) \, ds$. At that the density of tension force, i.e., the ratio $T(x)/S(x)$ is found to be constant.

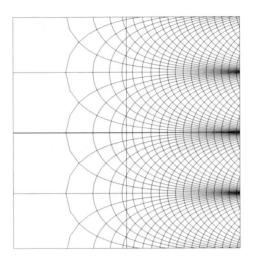

FIGURE 1.24

points $z = 0$ and $z = \infty$. A computer evaluates the principal branch of this function

$$w = \exp(\alpha \ln z), \qquad (1.7.37)$$

defined by the condition $-\pi < \arg z < \pi$. This branch is the single-valued analytic function in the z-plane with the cut along the negative real half-axis. The derivative of the function (1.7.37) $w' = \alpha z^{\alpha-1}$ does not vanish anywhere in the domain of definition, i.e., the mapping carried out by the function (1.7.37) is locally conformal everywhere.

At first let us consider the case of the real index $\alpha > 0$. The function (1.7.37) maps the point $z = \rho e^{i\varphi}$ into the point $w = R e^{i\Phi}$, where $R = \rho^\alpha$, $\Phi = \alpha \varphi$ and the polar net into the polar one. Let G be an angular domain

$$C < \arg z < D, \qquad (1.7.38)$$

where $-\pi \le C < D \le \pi$. The domain G is mapped by the function (1.7.37) onto the angular domain

$$C\alpha < \arg w < D\alpha.$$

If $0 < \alpha < 1$ then the function (1.7.37) is univalent in the whole domain of definition. If $\alpha > 1$ the function (1.7.37) to be univalent in the domain (1.7.38) it is necessary that $D\alpha - C\alpha < 2\pi$, i.e., the spread angle of the domain (1.7.38) $D - C$ should not exceed the value $2\pi/\alpha$. In particular, the half-plane $\operatorname{Re} z > 0$ is conformally mapped when $\alpha < 2$ onto the angular domain $-\pi\alpha/2 < \arg w < \pi\alpha/2$ with the spread angle $\pi\alpha$. If $\alpha < 1$ the spread angle is less than π. Such an angle (acute or obtuse) is called a *protruding* one. If $1 < \alpha < 2$ the spread angle is greater than π. Such an angle is called an *incoming* one. If $\alpha > 2$ the mapping of the half-plane onto an angular domain is not univalent.

The power function (1.7.37) is always used when it is necessary to map conformally the angular domain (1.7.38) onto a half-plane. For the purpose the power index α is chosen as $\alpha = \pi/(D - C)$.

Let us consider the mapping of the Cartesian net of lines carried out by the function (1.7.37). The straight line $x = C$ (or $\rho = C/\cos\varphi$ in polar coordinates) is mapped by the function (1.7.37) into the curve defined by the polar equation

$$R = \frac{C_1}{(\cos\Phi/\alpha)^\alpha} \,. \qquad (1.7.39)$$

In the case of $0 < \alpha < 1$ the curve (1.7.39) is called the *generalized hyperbola*. It represents equi-sided hyperbola if $\alpha = 1/2$ (see Subsection 1.7.3). In the case of $1 < \alpha < 2$ the curve (1.7.39) is called the *generalized parabola* (it presents a usual parabola if $\alpha = 2$). Any straight line not passing through the origin is mapped into the generalized parabola or hyperbola similar to (1.7.39) but disposed differently. The Cartesian net in the half-plane Re $z > 0$ is conformally mapped by the function (1.7.37) for $0 < \alpha < 2$ into two reciprocally-orthogonal families of generalized hyperbolas (or parabolas). If $\alpha > 2$ the curve (1.7.39) has a self-intersection. In this case the mapping of the half-plane is not univalent (conformal).

Example 1.9

The net of generalized parabolas representing the mapping of the Cartesian net by the function $w = z^{1.75}$ is depicted in Fig. 1.25. The isothermic net of this function is shown in Fig. 1.26. It represents two reciprocally-orthogonal families of the generalized hyperbolas.

Let us consider the mapping of the unipolar net of circumferences realized by the function (1.7.37). The circumference defined by the polar equation $\rho = C \cos\varphi$ is mapped by the function (1.7.37) into the curve defined by the equation

$$R = l \left(\cos\frac{\Phi}{\alpha}\right)^\alpha , \qquad (1.7.40)$$

where $l = C^\alpha$. In the case of $0 < \alpha < 1$ the curve (1.7.40) is called the *generalized lemniscata*. It represents a *Bernoulli's lemniscata* if $\alpha = 1/2$ (see Subsection 1.7.3). The finite domain bounded by the lemniscata has the protruding angle $\pi\alpha$ at the origin. In the case of $1 < \alpha < 2$ the curve (1.7.40) is called the *generalized cardioid* (if $\alpha = 2$ it represents a

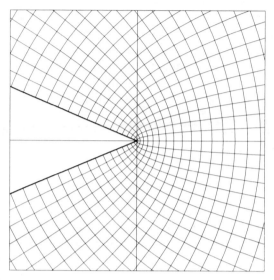

FIGURE 1.25

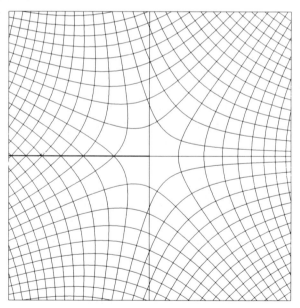

FIGURE 1.26

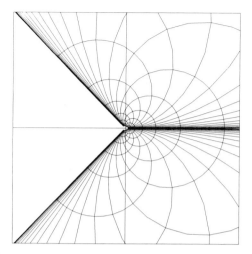

FIGURE 1.27

usual cardioid). The generalized cardioid bounds a domain having the incoming angle $\pi \alpha$.

The power function with the negative index $-\beta$ ($\beta > 0$) can be presented as the composition of two functions

$$\xi = \frac{1}{z}, \quad w = \xi^{\beta}.$$

The first of these functions (the inversion) maps the Cartesian net of the right half-plane into the unipolar net, the second one maps the circumferences of the unipolar net into the net of generalized lemniscatas or cardioids.

Example 1.10

The mapping of the Cartesian net of the half-plane Re $z > 0$ carried out by the function $w = 1/z^{1.5}$ is depicted in Fig. 1.27.

At last let us consider the case of the complex index α. In this case introduce the function (1.7.37) as the composition of three functions:

$$\eta = \ln z, \quad \xi = \alpha \eta, \quad w = e^{\xi}.$$

The first function maps the polar net of lines in the domain $-\pi < \arg z < \pi$ into the Cartesian net in the strip $-\pi < \operatorname{Im} \eta < \pi$. The second one maps the horizontal strip into the inclined one and finally the third function maps the rotated Cartesian net into a net of logarithmic spirals.

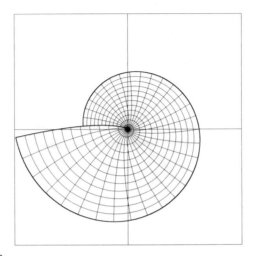

FIGURE 1.28

Example 1.11
The mapping of the polar net in the unit circle carried out by the function

$$w = z^{1/(1-ip)}$$

with $p = 0.15$ is presented in Fig. 1.28.

1.7.8   Hyperbolic and trigonometric functions

The hyperbolic functions of a complex variable are defined as

$$\cosh z = \frac{e^z + e^{-z}}{2}, \qquad \sinh z = \frac{e^z - e^{-z}}{2},$$

$$\tanh z = \frac{\sinh z}{\cosh z} = \frac{e^{2z} - 1}{e^{2z} + 1}, \qquad \coth z = \frac{\cosh z}{\sinh z} = \frac{e^{2z} + 1}{e^{2z} - 1},$$

$$\operatorname{sech} z = \frac{1}{\cosh z} = \frac{2}{e^z + e^{-z}}.$$

The functions $\cosh z$ and $\sinh z$ can be considered as the composition of the exponent and Zhukovskii function:

$$\xi = e^z, \qquad \cosh z = \tfrac{1}{2}\left(\xi + \tfrac{1}{\xi}\right),$$

$$\xi_1 = i\,e^z, \qquad \sinh z = \tfrac{1}{2i}\left(\xi_1 + \tfrac{1}{\xi_1}\right).$$

The functions $\tanh z$ and $\coth z$ are the compositions of the exponent and linear-fractional function:

$$\xi = e^{2z}, \quad \tanh z = \frac{\xi - 1}{\xi + 1}, \quad \coth z = \frac{\xi + 1}{\xi - 1}.$$

Finally, the function $w = \operatorname{sech} z$ can be considered as the composition of the exponent, Zhukovskii function, and the inversion:

$$\eta = e^z, \quad \xi = \frac{1}{2}\left(\eta + \frac{1}{\eta}\right), \quad w = \frac{1}{\xi}.$$

The functions $\cosh z$, $\sinh z$, $\operatorname{sech} z$ similarly to e^z have the pure imaginary period $2\pi i$. The functions $\tanh z$ and $\coth z$ are also periodic and like e^{2z} have the pure imaginary period πi.

The functions $\cosh z$ and $\sinh z$ are entire transcendental functions. The other hyperbolic functions are meromorphic ones with an infinite set of poles of the first order.

The trigonometric functions of a complex variable $\cos z$ and $\sin z$ are defined as analytic continuations of the functions $\cos x$ and $\sin x$ from the real axis. For the real values of the variable the Euler formula (1.1.2) yields

$$\cos x = \frac{e^{ix} + e^{-ix}}{2}, \quad \sin x = \frac{e^{ix} - e^{-ix}}{2i}.$$

Consider analytic continuations of these functions for any $z \in \mathcal{C}$

$$\cos z = \frac{e^{iz} + e^{-iz}}{2}, \quad \sin z = \frac{e^{iz} - e^{-iz}}{2i}. \tag{1.7.41}$$

The connection of trigonometric and hyperbolic functions is obvious:

$$\cos iz = \cosh z, \quad \sin iz = i \sinh z. \tag{1.7.42}$$

For the functions (1.7.41) all formulae (equalities) of the real analysis are valid (formulae of differentiation, addition theorems, formulae of double argument, etc.). In particular, addition theorems and formulae (1.7.42) yield

$$\cos(x + iy) = \cos x \cosh y - i \sin x \sinh y,$$
$$\sin(x + iy) = \sin x \cosh y + i \cos x \sinh y.$$

The relief map of the function $w = \cos z$ in the strip $-\pi < \operatorname{Re} z < \pi$ is shown in Fig. 1.29. Outside the strip the function is continued periodically.

The functions $\sin z$ and $\cos z$ are connected with each other by the relation $\sin z = \cos(z - \pi/2)$, i.e., by the linear transformation (translation) of the independent variable z. The hyperbolic functions $\sinh z$ and $\cosh z$ are also expressed through $\sin z$ and $\cos z$ with the help of the linear transformation of the independent variable (1.7.42) which represents the turn by the angle $\pi/2$. Consequently, for the investigation of conformal mappings carried out by the functions $\sin z$, $\cos z$, $\sinh z$, and $\cosh z$ it is enough to

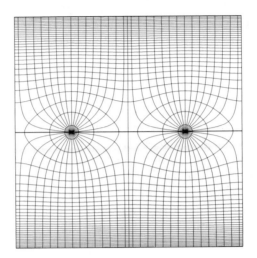

FIGURE 1.29

study one of them, for example, $\cos z$. The other mappings will be obtained from the investigated one by means of translation and (or) turn of the planes z and w by the angle $\pi/2$.

Let us determine the domain of univalence of the function $w = \cos z$. Critical points of the function are obtained from the condition $w' = -\sin z = 0$, whence $z = \pi n$. At all other points of the complex plane the function $w = \cos z$ is the locally univalent one. If the function acquires equal values at the points z_1 and z_2 then

$$\cos z_1 - \cos z_2 = 0$$

or

$$\sin \frac{z_1 - z_2}{2} \sin \frac{z_1 + z_2}{2} = 0 \,,$$

whence either $z_1 - z_2 = 2\pi n$ or $z_1 + z_2 = 2\pi n$ $(n \in \mathcal{Z})$. The function $\cos z$ is univalent in some domain if for any point z_1 in the domain the points $-z_1$, $2\pi - z_1$, $-2\pi - z_1$ do not lie in this domain. These conditions are fulfilled, for example, for the vertical strip $0 < x < \pi$, or for the inclined strip $0 < x - ay < \pi$.

To determine the image of the Cartesian net in the strip $0 < \operatorname{Re} z < \pi$ in the mapping $w = \cos z$ let us express $\cos z$ as the composition of the two functions:

$$\xi = e^{iz}, \quad w = \frac{1}{2} \left(\xi + \frac{1}{\xi} \right) \,.$$

The first function conformally maps the strip $0 < \operatorname{Re} z < \pi$ onto the half-plane $\operatorname{Im} \xi > 0$, the Cartesian net in the strip being mapped into the polar

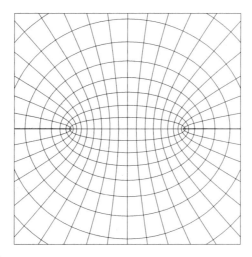

FIGURE 1.30

net in the half-plane. The second function (the Zhukovskii function) maps the upper half-plane ξ into the plane w with cuts along the rays of the real axis $(-\infty, -1)$ and $(1, \infty)$ (see Subsection 1.7.4). At that the polar net is mapped into the net of confocal ellipses and hyperbolas. This net represents coordinate line families of an elliptic coordinate system (see Subsection 1.8). The mapping of the Cartesian net in the strip $0 < \operatorname{Re} z < \pi$ realized by the function $w = \cos z$ is presented in Fig. 1.30.

The function $w = \sin z = \cos(z - \pi/2)$ is univalent, for example, in the strip $-\pi/2 < \operatorname{Re} z < \pi/2$. It conformally maps the Cartesian net in this strip onto the net of confocal ellipses and hyperbolas (Fig. 1.30).

The hyperbolic functions $w = \cosh z$ and $w = -i \sinh z$ are univalent correspondingly in the horizontal strips $0 < \operatorname{Im} z < \pi$ and $-\pi/2 < \operatorname{Im} z < \pi/2$. These functions also conformally maps the Cartesian net in their strips of univalence onto the net depicted in Fig. 1.30.

The function $w = \tan z = \sin z / \cos z$ is a meromorphic function with the infinite set of poles of the first order at the points $z = \pi(n + 1/2)$. The function $\tan z$ is the periodic one with the real period π. It is connected with the hyperbolic tangent by the relation

$$\tan z = -i \tanh iz .$$

The relief map of the function $w = \tan z$ in the strip $-\pi/2 < \operatorname{Re} z < \pi/2$ is depicted in Fig. 1.31. The function is continued periodically beyond the boundaries of the named strip.

The function $w = \tan z$ can be represented as a composition of the two

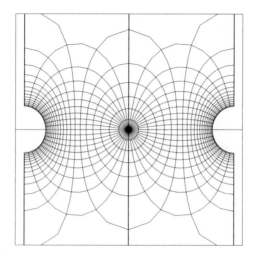

FIGURE 1.31

functions

$$\xi = e^{2iz}, \quad w = -i\frac{\xi - 1}{\xi + 1}. \tag{1.7.43}$$

The exponent conformally maps the strip $-\pi/2 < \operatorname{Re} z < \pi/2$ onto the ξ-plane with the cut along the negative real half-axis, the Cartesian net in the strip being mapped onto the polar net in the ξ-plane. The linear-fractional function maps the plane with the cut along the ray onto the w-plane with two cuts along the rays of the imaginary axis $(-i\infty, -i)$ and $(i, i\infty)$. At that the polar net of lines is conformally mapped onto the bipolar net with the poles at the points $\pm i$.

The mapping of the Cartesian net in the strip $-\pi/2 < \operatorname{Re} z < \pi/2$ realized by the function $w = \tan z$ is presented in Fig. 1.32. Vertical straight lines $\operatorname{Re} z = C$ are mapped onto arcs of Steiner's circumferences passing through the points i and $-i$ and segments of horizontal straight lines are mapped onto the Apollonian circumferences.

Let us consider what line the function $w = \tan z$ maps the inclined straight line $y - ax = C$ onto. Representing $\tan z$ as the composition of functions (1.7.43) one obtains the straight line in the z-plane is mapped onto the logarithmic spiral in the ξ-plane, which is mapped onto loxodrome in the w-plane under the linear-fractional transformation.

Example 1.12

The function $w = \tan(z/(1 - ip))$ is univalent in the strip $-\pi/2 < \operatorname{Re} z < \pi/2$ and conformally maps the Cartesian net onto the net of reciprocally orthogonal families of *loxodromes*. The mapping of the Cartesian net carried

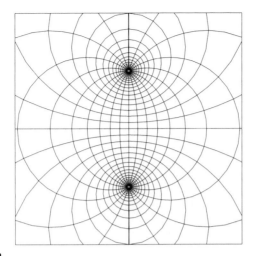

FIGURE 1.32

out by the function

$$w = \tan \frac{z}{1 - 0.477i}$$

is shown in Fig. 1.33.

Finally consider the mapping realized by the function $w = \sec z = 1/\cos z$. Like $\cos z$ this function is univalent in the strip $0 < \operatorname{Re} z < \pi$. It can be presented as a composition of the two functions

$$\xi = \cos z, \quad w = 1/\xi.$$

As was shown above, the function $\xi = \cos z$ conformally maps the strip $0 < \operatorname{Re} z < \pi$ onto the plane ξ with the cuts along the rays of the real axis, the Cartesian net being mapped into the net of confocal ellipses and hyperbolas. The second transformation represents the inversion with respect to the common center of ellipses and hyperbolas.

The inversion of the conic curves with respect to their center gives the curves called the *Booth's lemniscatas*. More accurately, by the inversion of ellipses with respect to the center one obtains smooth closed Jordan curves, called *elliptic Booth's lemniscatas*. The inversion of the hyperbola consisting of two branches gives the closed curve with the self-intersection point at the origin. This curve is called *hyperbolic Booth's lemniscata.*¶

Therefore, the Cartesian net in the strip $0 < \operatorname{Re} z < \pi$ is conformally mapped by the function $w = \sec z$ onto two reciprocally-orthogonal families

---

¶Bernoulli's lemniscata considered above is the particular case of Booth's lemniscata, it is an inversion of equi-sided hyperbola.

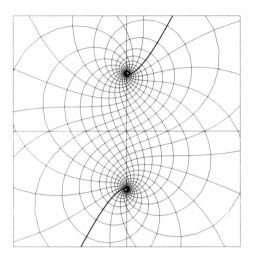

FIGURE 1.33

of lines, one of which consists of an elliptic, and the second one consists of hyperbolic Booth's lemniscatas. This net of lines is depicted in Fig. 1.34.

Exercises

1. With the help of the program CONFORM create the isothermic net of the function

$$w = \sin z \,.$$

2. Create the mapping of the strip $-\pi/2 < \operatorname{Im} z < \pi/2$ by the function

$$w = \sinh \frac{z}{1 - ip}$$

   for the values of the parameter $p = -0.5, 0, 0.6$.

1.7.9 Inverse trigonometric and hyperbolic functions

The function $w = \operatorname{Arccos} z$ is defined as a solution of the equation $z = \cos w$. Represented as

$$z = \frac{e^{iw} + e^{-iw}}{2}$$

it is reduced to the quadratic equation with respect to the function e^{iw}. The equation has the infinite set of solutions

$$w = -i \operatorname{Ln}\left(z \pm i\sqrt{1 - z^2}\right).$$

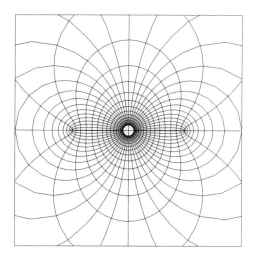

FIGURE 1.34

The function $\operatorname{Arccos} z$ is an infinite-valued one with the branch points $z = \pm i$. To separate its single-valued branch a cut should be made connecting the branch points. The cut is usually made along the rays of the real axis $(-\infty, -1)$ and $1, \infty)$. As the principal value of the multiply-valued function

$$\arccos z = -i \ln \left(z + i\sqrt{1 - z^2} \right) \qquad (1.7.44)$$

is chosen.

The function $w = \arccos z$ coincides with the real function $\arccos x$ on the interval $(-1, 1)$ of the real axis.$^{\|}$ Therefore, it represents the analytic continuation of the function $\arccos x$ from the interval $(-1, 1)$ onto the complex plane with the cuts along the rays of the real axis.

The function $w = \cos z$, as was shown in the previous Subsection, conformally maps the strips $0 < \operatorname{Re} z < \pi$ onto the plane w with the cuts along the rays of the real axis. The inverse function $w = \arccos z$ performs the inverse mapping of the z-plane with the cuts along the rays onto the strip $0 < \operatorname{Re} z < \pi$. The right cut ($z = x > 1$) is mapped by the function (1.7.44) onto the imaginary axis $\operatorname{Re} w = 0$ and the left cut is mapped into the straight line $\operatorname{Re} w = \pi$. One can follow the correspondence of boundaries in the mapping analytically, but it is much simpler to build the mapping using a computer with the help of the program CONFORM (see below the Exercise 1).

$^{\|}$Indeed, when $z = x = \cos t$ (where $0 < t < \pi$) one has $z + i\sqrt{1 - z^2} = \cos t + i \sin t = e^{it}$ and $w = -i \ln \left(e^{it} \right) = t = \arccos x$.

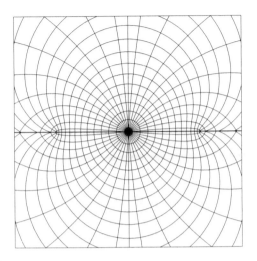

FIGURE 1.35

The function $w = \arcsin z$ can be defined in a similar way as

$$w = -i \ln \left(\sqrt{1 - z^2} + iz \right).$$ (1.7.45)

The domain of definition of this function as well as of the function (1.7.44) is the z-plane with the cuts along the rays of the real axis. In the interval $(-1, 1)$ the function (1.7.4) assumes the real values $w = \arcsin x$ and, therefore, it is the analytic continuation of the function $\arcsin x$ from the interval $(-1, 1)$ onto the complex plane with cuts along the rays. In the interval $(-1, 1)$ the real functions $\arcsin x$ and $\arccos x$ are connected by the relation

$$\arcsin x + \arccos x = \frac{\pi}{2}.$$

Due to the theorem on the analytic continuation of equalities this formula is valid in all the domain of the analyticity of the functions (1.7.44), (1.7.45), so

$$\arcsin z = \frac{\pi}{2} - \arccos z.$$

This formula yields that the function $w = \arcsin z$ conformally maps the z-plane with the cuts along the rays of the real axis onto the strip $-\pi/2 < \operatorname{Re} w < \pi/2$.

The relief map of the function $\arcsin z$ is presented in Fig. 1.35.

The function $w = \operatorname{Arctan} z$ is defined as the solution of the equation

$$z = -i \frac{e^{2iw} - 1}{e^{2iw} + 1}.$$

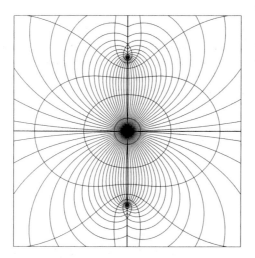

FIGURE 1.36

This equation has the infinite set of solutions

$$w = -\frac{i}{2} \operatorname{Ln} \frac{1+iz}{1-iz}.$$

The function $\operatorname{Arctan} z$ is an infinite-valued one with the branch points $z = \pm i$. To separate the single-valued branch a cut should be made connecting the branch points. The cut is usually made along the rays of the imaginary axis $(-i\infty, -i)$ and $(i, i\infty)$. As the principal branch of the multiply-valued function

$$\arctan z = -\frac{i}{2} \ln \frac{1+iz}{1-iz} \qquad (1.7.46)$$

is chosen.

The function (1.7.46) coincides with the real function $\arctan x$ on the real axis (when $z = x$).**

Therefore, it is the analytic continuation of the function $\arctan x$ from the real axis onto the complex plane with the cuts along the rays of the imaginary axis.

The relief map of the function $\arctan z$ is presented in Fig. 1.36.

The function $w = \arctan z$ can be considered as the composition of two functions:

$$\xi = \frac{1+iz}{1-iz}, \quad w = -\frac{i}{2} \ln \xi.$$

**Indeed, let $z = x = \tan t$ (where $-\pi/2 < t < \pi/2$). Then $w = -i/2 \ln \left(e^{i2t}\right) = t = \arctan x$.

The first linear-fractional function conformally maps the z-plane with the cuts along the rays of the imaginary axis onto the ξ-plane with the cut along the negative real half-axis. The second function maps the ξ-plane with the cut onto the strip $-\pi/2 < \operatorname{Re} w < \pi/2$. The reader is suggested to create by himself the mapping of the polar net by this function (the Exercise 4).

The inverse hyperbolic functions are also infinite-valued functions. Their principal values are defined as:

$$\operatorname{arsh} z = \ln\left(z + \sqrt{z^2 + 1}\right),$$

$$\operatorname{arch} z = \ln\left(z + \sqrt{z^2 - 1}\right),$$

$$\operatorname{arth} z = \frac{1}{2} \ln \frac{1+z}{1-z}.$$

The inverse hyperbolic functions are rather simply expressed through the inverse trigonometric ones:

$$\operatorname{arsh} z = -i \, \arcsin(iz),$$

$$\operatorname{arch} z = i \, \operatorname{arccos} z,$$

$$\operatorname{arth} z = -i \, \arctan(iz).$$

Exercises

1. Create the mapping of the polar net in the upper half-plane carried out by the function
$$w = \operatorname{arccos} z$$
(consider the sector $0 < \rho < 3$, $0.001 < \varphi < 3.141$). Interpret the obtained figure as the relief map of the function $\cos z$ in some domain and compare it with the Fig. 1.29.

2. Create the mapping of the Cartesian net in the upper half-plane carried out by the function
$$w = \arcsin z$$
(consider the rectangle $-3 < x < 3$, $0.001 < y < 3$). Interpret the created image as the isothermic net of the function $\sin z$.

3. Create the mapping of the rectangle realized by the function
$$w = \arcsin(\sin z).$$
Make sure that if the rectangle is located inside the strip $-\pi/2 < \operatorname{Re} z < \pi/2$ it is mapped onto itself, i.e., for such z the equality $\arcsin(\sin z) = z$ is true. Show that the square $|x| < 3$, $|y| < 3$ is not mapped onto itself.

4. Create the mapping of the polar net inside the unit circle $|z| < 1$ realized by the function
$$w = \arctan z$$

and interpret the created net of lines as the relief map of the function $\tan z$ in some domain. Compare the result with the Fig. 1.31.

5. Create the mapping of the polar net carried out by the function

$$w = \arctan(\tan z)$$

(consider the circle $0 < \rho < 3.14$, $-3.14 < \varphi < 3.14$). Make sure that the circle $|z| < \pi/2$ is mapped onto itself, i.e., in this domain $\arctan(\tan z) = z$ (outside the strip $-\pi/2 < \operatorname{Re} z < \pi/2$ this equality is not true).

1.8   Isothermic Coordinates. Differential Operators in the Isothermic Coordinates

Let the analytic function $w = u + iv = f(z)$ defined in the domain G conformally map the domain G onto the domain D. In this case the variables u and v uniquely determining the position of the point $z = x + iy$ in the domain G represent the isothermic coordinates of the point z (see Section 1.5). The inverse expression of the Cartesian coordinates x, y through the isothermic ones has the form

$$z = x + iy = F(w),$$

where $F = f^{-1}$ is the inverse function. The arc element length $ds = |dz|$ is expressed in the isothermic coordinates as:

$$ds = |F'(w)|\,|dw|. \tag{1.8.1}$$

Designating $H = |F'(w)|$ let us write the expression (1.8.1) as

$$ds^2 = H^2(du^2 + dv^2). \tag{1.8.2}$$

The value H is called the Lame coefficient.* To express differential operators in isothermic coordinates the formal derivatives apparatus is convenient to be used. Considering $z = x + iy$ and $\bar{z} = x - iy$ as independent variables one obtains

$$\varphi(x, y) = \varphi\left(\frac{z + \bar{z}}{2}, \frac{z - \bar{z}}{2i}\right) = \Phi(z, \bar{z}),$$

$$\frac{\partial \varphi}{\partial x} = \frac{\partial \Phi}{\partial z} + \frac{\partial \Phi}{\partial \bar{z}}, \qquad \frac{\partial \varphi}{\partial y} = i\left(\frac{\partial \Phi}{\partial z} - \frac{\partial \Phi}{\partial \bar{z}}\right). \tag{1.8.3}$$

---

*In arbitrary orthogonal curvilinear coordinates the arc element length is expressed as

$$ds^2 = H_1^2\,du^2 + H_2^2\,dv^2,$$

where H_1, H_2 are Lame coefficients. In the case of isothermic coordinates $H_1 = H_2$.

The expressions (1.8.3) are called *formal derivatives* of the function $\varphi(x, y)$. By combining them one obtains inverse relations

$$2\frac{\partial \Phi}{\partial z} = \frac{\partial \varphi}{\partial x} - i\frac{\partial \varphi}{\partial y}, \quad 2\frac{\partial \Phi}{\partial \bar{z}} = \frac{\partial \varphi}{\partial x} + i\frac{\partial \varphi}{\partial y}. \tag{1.8.4}$$

Using formal derivatives the conditions of analyticity of the function $w = f(z)$ (1.2.1) take especially simple appearance

$$\frac{\partial f}{\partial \bar{z}} = 0.$$

The operator of the first order

$$V = (\nabla \varphi)^2 = \left(\frac{\partial \varphi}{\partial x}\right)^2 + \left(\frac{\partial \varphi}{\partial y}\right)^2$$

and the Laplace operator

$$W = \nabla^2 \varphi = \frac{\partial^2 \varphi}{\partial x^2} + \frac{\partial^2 \varphi}{\partial y^2}$$

are called basic differential operators for the real-valued function $\varphi(x, y)$. They are rather simply expressed through the formal derivatives:

$$V = 4\left|\frac{\partial \Psi}{\partial z}\right|^2, \quad W = 4\frac{\partial^2 \Psi}{\partial z \, \partial \bar{z}}. \tag{1.8.5}$$

If the conversion from the Cartesian coordinates x, y to the isothermic coordinates u, v is carried out with the help of the analytic function $w = f(z)$, the formal derivatives are transformed according to the rule

$$\frac{\partial}{\partial z} = f'(z)\frac{\partial}{\partial w}, \quad \frac{\partial}{\partial \bar{z}} = \overline{f'(z)}\frac{\partial}{\partial \bar{w}}.$$

Making the variable substitution $w = f(z)$ in expressions (1.8.5), returning to the ordinary derivatives $\partial \varphi / \partial u$ and $\partial \varphi / \partial v$ and taking into account

$$|f'(z)| = \frac{1}{|F'(w)|} = \frac{1}{H},$$

one obtains the expression of the basic differential operators in the isothermic coordinates u, v:

$$(\nabla \varphi)^2 = \frac{1}{H^2}\left[\left(\frac{\partial \varphi}{\partial u}\right)^2 + \left(\frac{\partial \varphi}{\partial v}\right)^2\right], \tag{1.8.6}$$

$$\nabla^2 \varphi = \frac{1}{H^2}\left(\frac{\partial^2 \varphi}{\partial u^2} + \frac{\partial^2 \varphi}{\partial v^2}\right). \tag{1.8.7}$$

The important conclusion about invariance of harmonic function in the conformal transformation of variables follows from the formula of Laplace operator transformation (1.8.7). If the function $\varphi(x, y)$ satisfies the Laplace equation $\nabla^2 \varphi = 0$, it also satisfies the Laplace equation in the isothermic coordinates u, v:

$$\frac{\partial^2 \varphi}{\partial u^2} + \frac{\partial^2 \varphi}{\partial v^2} = 0.$$

In mathematical physics (see Reference[101], for example) the following orthogonal curvilinear coordinates in the plane are often used:

(a) Polar coordinates. The polar coordinates ρ, φ themselves (where $z = \rho e^{i\varphi}$) are not isothermic, however the variables $u = \ln \rho$ and $v = \varphi$ are isothermic coordinates determined by the analytic function $w = \ln z$, $z = e^w$ and conditions $-\infty < u < \infty$, $-\pi < v < \pi$. The Lame coefficient of this coordinate system equals $H = |e^w| = e^u = \rho$. The formulae (1.8.2), (1.8.6), and (1.8.7) take the form

$$ds^2 = e^{2u} (du^2 + dv^2) = d\rho^2 + \rho^2 d\varphi^2,$$

$$(\nabla U)^2 = e^{-2u} \left[\left(\frac{\partial U}{\partial u} \right)^2 + \left(\frac{\partial U}{\partial v} \right)^2 \right] = \left(\frac{\partial U}{\partial \rho} \right)^2 + \frac{1}{\rho^2} \left(\frac{\partial U}{\partial \varphi} \right)^2,$$

$$\nabla^2 U = e^{-2u} \left(\frac{\partial^2 U}{\partial u^2} + \frac{\partial^2 U}{\partial v^2} \right) = \frac{\partial^2 U}{\partial \rho^2} + \frac{1}{\rho} \frac{\partial U}{\partial \rho} + \frac{1}{\rho^2} \frac{\partial^2 U}{\partial \varphi^2}.$$

(b) The parabolic coordinates represent the isothermic coordinates defined by the relation $z = 1/2 \, w^2$ and conditions $u > 0$, $-\infty < v < \infty$. The Lame coefficient of this coordinate system equals $H^2 = |w|^2 = u^2 + v^2$. The coordinate net of lines is depicted in Fig. 1.37.

(c) The elliptic coordinates are determined by the analytic function $z = \cosh w$ and the conditions $u > 0$, $-\pi < v < \pi$. The Lame coefficient of this coordinate system equals $H^2 = |\sinh w|^2 = \cosh^2 u - \cos^2 v$. The coordinate net of lines of the elliptic coordinates is depicted in Fig. 1.38.

(d) The bipolar coordinates in the plane are determined by the relation $z = \tanh(w/2)$ and conditions $-\pi < v < \pi$. The coordinate lines net of this coordinate system is depicted in Fig. 1.39. The Lame coefficient for the bipolar coordinates equals

$$H = \frac{1}{\left| 2 \cosh^2 \frac{w}{2} \right|} = \frac{1}{\cosh u + \cos v}.$$

(e) The unipolar (or degenerate bipolar) coordinates are introduced by the relation $z = 1/w = 1/(u + iv)$, in other words, the conversion to the unipolar coordinates is carried out by the inversion transforma-

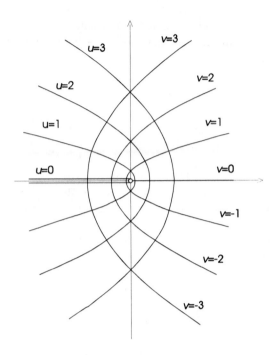

FIGURE 1.37

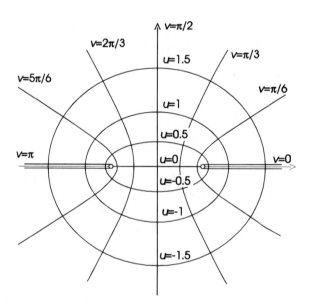

FIGURE 1.38

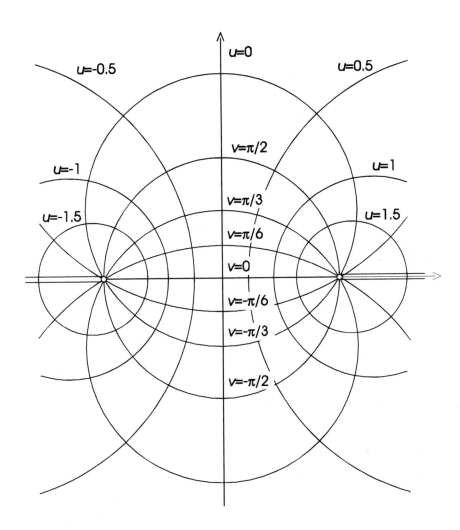

FIGURE 1.39

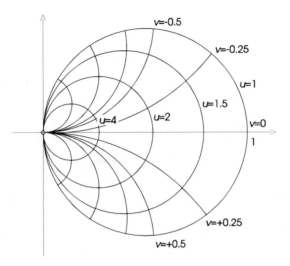

FIGURE 1.40

tion. The Lame coefficient of this coordinate system equals

$$H = \frac{1}{|w|^2} = \frac{1}{u^2 + v^2}.$$

The coordinate net of lines of the unipolar coordinates in the circle $|z - 1/2| < 1/2$ is presented in Fig. 1.40. This net of lines is widely used in radio engineering where it is called the Smith's diagram.[145]

1.9 Mappings of Lunes

The *circular lune* (or circular two-sided polygon) is the singly connected domain of the extended complex plane bounded by two arcs of circumferences (in particular, such arcs can be segments of straight lines). Such domains can be conformally mapped onto canonical domains with the composition of elementary functions.

In the simplest cases both of the boundary arcs are parts of straight lines. In these cases the domain is degenerated into the straight linear lune, being an angular domain, a strip or an exterior of the straight line segment. In the current Section conformal mappings of nondegenerate circular lunes are considered.

The circumferences that bound a circular lune can be intersecting or tangent to each other. In the case of intersecting circumferences the points of circumferences intersection can be placed into the points $w = \pm 1$ by making the linear transformation. In the case of tangent circumferences we shall suppose the circumferences to be tangent to the real axis at the origin, the diameter of a certain circumference being equal to 1. This also can be obtained with the help of linear transformation.

1.9.1 Curvilinear half-planes

Let D be the upper half-plane Im $w > 0$ with the circle sector deleted (or added), which boundary passes through the points $w = \pm 1$. Let angles of the domain D at the points ± 1 equal to πp (if $0 < p < 1$ the angles are protruding ones, if $1 < p < 2$ they are incoming ones).

The linear-fractional transformation

$$t = \frac{w - 1}{w + 1}$$

conformally maps the domain D onto the angular domain $0 < \arg t < \pi p$, further the power function $s = t^{1/p}$ maps the angle onto the upper half-plane Im $s > 0$. The point $w = \infty$ is mapped into the point $s = 1$. To obtain the canonical mapping of the domain D onto the upper half-plane Im $z > 0$ the linear-fractional transformation should be carried out once more:

$$z = \frac{1 + s}{1 - s}.$$

The inverse mapping of the half-plane Im $z > 0$ onto the domain D is realized by the composition of the functions

$$s = \frac{z - 1}{z + 1}, \quad t = s^p, \quad w = \frac{1 + t}{1 - t}, \tag{1.9.1}$$

which is conveniently rewritten in the implicit form

$$\frac{w - 1}{w + 1} = \left(\frac{z - 1}{z + 1} \right)^p. \tag{1.9.2}$$

The function (1.9.2) is called the *generalized Zhukovskii function*. When $p = 2$ it represents the ordinary Zhukovskii function, when $p = 1/2$ it is inverse to the Zhukovskii function, when $p = 1$ it is the identity transformation.

The basic properties of the generalized Zhukovskii function are as follows:

1. The function (1.9.2) is analytic everywhere except for the branch points $z = \pm 1$. The principal branch of the function is analytic in the complex plane with the cut along the interval $(-1, 1)$.

2. When $0 < p < 1$ the function (1.9.2) is univalent inside all the domain of definition. When $1 < p < 2$ it is univalent outside any circumference passing through the points $z = \pm 1$, for example, outside the unit circumference or inside the upper half-plane Im $z > 0$.

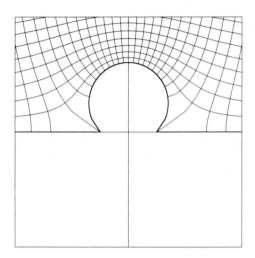

FIGURE 1.41

3. The function inverse to the generalized Zhukovskii function with the parameter p is the generalized Zhukovskii function with the parameter $1/p$.

4. The function (1.9.2) maps the bipolar net of lines with poles at the points $z = \pm 1$ onto the bipolar net with poles at the same points. Indeed, the first of the three transformations (1.9.1) maps the bipolar net onto the polar one, in the second transformation $t = s^p$ the polar net is mapped onto itself, the third (linear-fractional) transformation again maps the polar net onto the bipolar one. The mapping (1.9.2) is isogonal at all points except for $z = \pm 1$. The angles between arcs of Steiner's circumferences passing through the points $z = \pm 1$ are magnified in the mapping (1.9.2) by a factor of p.

Example 1.13

The mapping of the Cartesian net in the upper half-plane by the function

$$s = \frac{z-1}{z+1}, \quad t = s^{1/4}, \quad w = \frac{1+t}{1-t}$$

is depicted in Fig. 1.41.

Let the domain D be the upper half-plane $\operatorname{Im} w > 0$ with the circle $|w - i/2| > 1/2$ deleted. The inversion $t = -\pi/w$ maps this domain onto the strip $0 < \operatorname{Im} t < \pi$. Further, the function $s = e^t$ maps the strip onto the upper half-plane $\operatorname{Im} s > 0$, the point $w = \infty$ being mapped into the

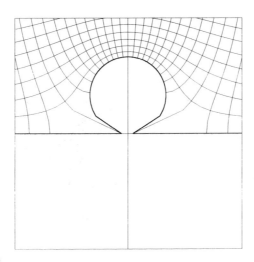

FIGURE 1.42

point $s = 1$. Finally, the linear-fractional function $z = 1/(1 - s)$ maps the upper half-plane onto itself, the point $s = 1$ being mapped into the point at infinity.

The inverse mapping of the upper half-plane Im $z > 0$ onto the curvilinear half-plane D is realized by the function

$$w = -\frac{\pi}{\ln(1 - 1/z)}.$$

The mapping of the Cartesian net carried out by this function is depicted in Fig. 1.42.

1.9.2   Exteriors of the finite contours

Let the domain D be the exterior of the contour consisting of two arcs of circumferences with the ends at the points $w = \pm 1$. Let the angles of the domain D at the points ± 1 equal to πp (where $0 < p < 2$). To construct the conformal mapping of the domain D onto the exterior of a circle one can use properties of the generalized Zhukovskii function (1.9.2). As was established in the previous Subsection the generalized Zhukovskii function with the parameter $1/p$ conformally maps the bipolar net of lines onto the bipolar net, angles between arcs of Steiner's circumferences at the points ± 1 being decreased by a factor of p. The angle between the boundary arcs of the domain D becomes equal to π in this transformation, i.e., the boundary of the domain D is mapped onto the circumference passing through the points $z = \pm 1$ and the domain D itself is mapped onto the exterior of the circle.

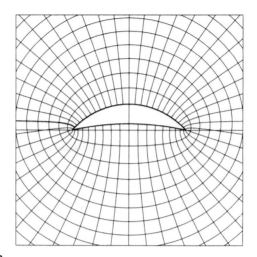

FIGURE 1.43

The inverse mapping of the exterior of the circle onto the domain D is carried out by the generalized Zhukovskii function with the parameter p.

Example 1.14

The mapping of the polar net in the exterior of the circle by the generalized Zhukovskii function is depicted in Fig. 1.43. It can be constructed with the help of the program CONFORM using the function:

$$a = 0.3, \quad p = 1.8$$
$$r = \frac{z + i \sin a}{\cos a}$$
$$s = \frac{r - 1}{r + 1}$$
$$t = s^p$$
$$w = \frac{1 + t}{1 - t}$$

Let the domain D be the infinite w-plane without two touching circles:

$$\left| w + \frac{i}{2} \right| > \frac{1}{2}, \quad \left| w - \frac{i}{2p} \right| > \frac{1}{2p}, \quad p > 0.$$

The mapping of this domain onto the exterior of the unit circle $|z| > 1$ is

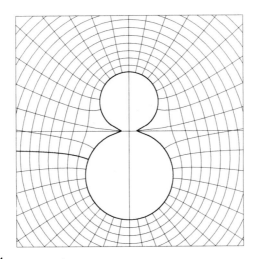

FIGURE 1.44

realized by the sequence of transformations

$$t = -\frac{1}{w}, \quad s = \pi\,\frac{t+i}{p+1}, \quad r_0 = \exp\left(\frac{\pi i}{p+1}\right),$$

$$r = e^s, \quad z = \frac{r - \overline{r_0}}{r - r_0}\,e^{-i\alpha},$$

where α is an arbitrary real number.

The inverse mapping of the exterior of the unit circle onto the domain D is carried out by the sequence of transformations

$$r_0 = \exp\left(\frac{i\pi}{p+1}\right), \quad r = \frac{z\,e^{i\alpha}\,r_0 - \overline{r_0}}{z\,e^{i\alpha} - 1}, \quad s = \ln r,$$

$$t = \frac{s(p+1)}{\pi} - i, \quad w = -\frac{1}{t} \tag{1.9.3}$$

To fulfill the condition $w'(\infty) > 0$ one should take $e^{i\alpha} = -i/r_0$.

Example 1.15
The mapping of the exterior of the unit circle realized by the function (1.9.3) when $p = 1.5$ is depicted in Fig. 1.44.

1.9.3   Finite lunes

Let the finite domain D be bounded by two arcs of the circumferences that intersect at the points $w = \pm 1$ and make the angle πp at them. The

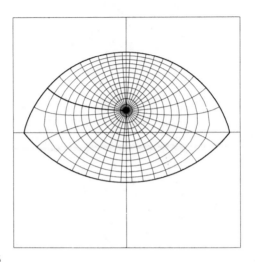

FIGURE 1.45

linear-fractional transformation

$$t = \frac{1 + w}{1 - w}$$

conformally maps the domain D onto an angular domain and the power
function $s = t^{1/p}$ maps it onto a half-plane. Let the normal to the boundary
straight line of the half-plane form the angle a with the real axis. Then the
transformation

$$r = \frac{s - 1}{s + 1}, \quad z = r \cos a - i \sin a$$

maps this half-plane onto the circle $|z| < 1$.

The inverse mapping of the unit circle onto the domain D is carried out
by the sequence of transformations

$$r = \frac{z + i \sin a}{\cos a}, \quad s = \frac{1 + r}{1 - r},$$

$$t = s^p, \quad w = \frac{t - 1}{t + 1}. \tag{1.9.4}$$

Example 1.16
The mapping of the polar net inside the circle $|z| < 1$ realized by the
function (1.9.4) with $a = 0.3$, $p = 0.7$ is presented in Fig. 1.45.

Let D be the domain enclosed between two circumferences that touch the real axis at the point $w = 0$:

$$\left| w - \frac{i}{2} \right| < \frac{1}{2}, \quad \left| w - \frac{i}{2p} \right| > \frac{1}{2p},$$

where $p > 1$. The linear-fractional transformation

$$t = -\frac{1}{w}, \quad s = \pi \frac{t - i}{p - 1}$$

conformally maps the domain D onto the strip $0 < \operatorname{Im} s < \pi$ and the exponent $r = e^s$ maps it onto the half-plane $\operatorname{Im} r > 0$. Finally, the linear-fractional function (1.7.16) maps the upper half-plane onto the unit circle, for example:

$$z = \frac{r - i}{r + i}.$$

The inverse mapping of the unit circle $|z| < 1$ onto the domain D is realized by the sequence of transformations

$$r = i \frac{1 + z}{1 - z}, \quad s = \ln r,$$

$$t = \frac{s(p - 1)}{\pi} + i, \quad w = -\frac{1}{t}. \tag{1.9.5}$$

Example 1.17
The mapping of the polar net inside the circle $|z| < 1$ realized by the function (1.9.5) with $p = 4$ is presented in Fig. 1.46. Essential distortions of the image near the origin should be mentioned. The even net used by the program CONFORM is not thick enough in the neighborhood of singularity of the function (1.9.5).

1.10   The Construction of Conformal Mappings with the Help of the Boundary Correspondence Principle

Let a closed curve L be given in the complex plane $w = u + iv$. At first let us assume for simplicity that the curve L is a Jordan line determined by the parametric equation

$$u = u(t), \quad v = v(t),$$

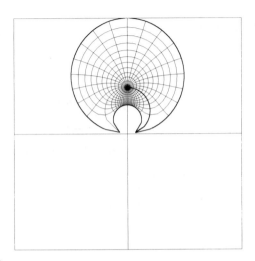

FIGURE 1.46

where $-\pi < t < \pi$ and $u(t)$, $v(t)$ are analytic real-valued functions of the real variable t.*

Let us further suppose that the curve L is smooth, i.e., $\dot{u}^2(t) + \dot{v}^2(t) \neq 0$ and the growth of the parameter t corresponds to the traversal of the curve L in a counterclockwise direction.

The function $w = u(t) + iv(t)$ carries out the reciprocal one-to-one mapping of the unit circumference Γ defined by the equation $z = \exp(it)$ onto the curve L. This function allows the analytic continuation onto some neighborhood of the circumference Γ — the annulus $1 - \varepsilon < |z| < 1 + \varepsilon$. Analytically continued function $w(z)$ conformally maps this annulus onto some neighborhood of the curve L.

Jordan curve L partitions the extended complex plane w into two singly connected domains — the finite domain D and infinite domain D^*. The function $w(z)$ can turn to have an analytic continuation inside the circle $|z| < 1.$[†] Then according to the Theorem 1.11 this function conformally maps the circle onto the domain D.

If the function $w(z)$ is analytic in the exterior of the circle $|z| > 1$ everywhere except for the single point, in which the function has the first order pole, it conformally maps the exterior of the circle onto the domain D^* in accordance with the Theorem 1.12.

In the general case it is not necessary to demand the curve L to be a

---

*A real-valued function given in the segment $a < t < b$ is called analytic if it is expandable into a power series by powers of $(t - t_0)$ in the neighborhood of each point t_0 of the segment.

[†] Of course, the possibility of an analytic continuation is determined not only by the shape of the curve L, but also by the means of its parametrization.

smooth bounded Jordan line. If the function $w(z)$ is analytic inside the circle $|w| < 1$, continuous in the closed circle, maps the circumference $|z| = 1$ reciprocally one-to-one onto the curve L — the boundary of some finite singly connected domain D, then it conformally maps the circle $|z| < 1$ onto the domain D. If the function $w(z)$ is analytic everywhere outside the unit circle, continuous up to the boundary of the circle, has the pole of the first order at the point at infinity and maps reciprocally one-to-one the unit circumference $|z| = 1$ onto the boundary L of some infinite singly connected domain D^*, for which the point at infinity is the interior one, then $w = w(z)$ conformally maps the domain $|z| > 1$ onto the domain D^*.

Let us consider how to use the boundaries correspondence principle to construct conformal mappings of domains bounded by some important curves on the basis of their parametric equations.

1.10.1 The ellipse

The ellipse with the half-axis a, b is defined by the parametric equation

$$u = a\cos t, \quad v = b\sin t,$$

or

$$w = \frac{a}{2}\left(e^{it} + e^{-it}\right) + \frac{b}{2}\left(e^{it} - e^{-it}\right).$$

Designating $z = \exp(it)$ the reciprocal one-to-one mapping of the unit circumference onto the ellipse is obtained:

$$w = \frac{a+b}{2} z + \frac{a-b}{2z}. \qquad (1.10.1)$$

The analytic continuation of the function (1.10.1) is called the *modified Zhukovskii function*. Outside the unit circle the generalized Zhukovskii function is analytic, it has the first order pole at the point at infinity and conformally maps the exterior of the circle onto the exterior of the ellipse.

Inside the unit circle the function (1.10.1) is not analytic; it has the pole at the origin. As well it is not univalent there. Indeed, the necessary condition of the univalence

$$w' = \frac{a+b}{2} - \frac{a-b}{2z^2} \neq 0$$

is violated at the points $z = \pm\sqrt{(a-b)/(a+b)}$. However in the annulus $\sqrt{(a-b)/(a+b)} < |z| < 1$ the function (1.10.1) is univalent. It conformally maps the annulus onto the doubly connected domain being the interior of the ellipse with the deleted segment between the ellipse foci $(-c, c)$, where $c = \sqrt{a^2 - b^2}$.

The interior of the ellipse (without cuts) can not be conformally mapped onto the circle by elementary functions. The conformal mapping of the circle onto the interior of the ellipse carried out with the help of elliptic integrals is presented in Catalog 1 (Domain 16).

1.10.2   Cassini's oval and Bernoulli's lemniscata

A set of points in the w-plane with the constant product of distances to the two given points (foci) is called the *Cassini's oval* or cassinian. If the foci are at the points of the real axis c and $-c$, the equation of the Cassini's oval has the form

$$|w - c| \cdot |w + c| = a^2$$

or

$$|w^2 - c^2| = a^2 .$$

When $a > c$ the Cassini's oval is a smooth Jordan curve (*single-contour oval*). When $a = c$ the cassinian is the *Bernoulli's lemniscata*, which is non-Jordan curve with the self-intersection point at the origin. When $a < c$ it represents a non-connected curve consisting of two ovals (*two-contour Cassini's oval*).

Singly-contour Cassini's oval $(a > c)$ is parametrized in the following way:

$$w^2 - c^2 = a^2 e^{2it} ,$$

whence

$$w = \sqrt{c^2 + a^2 e^{2it}} = e^{it} \sqrt{a^2 + c^2 e^{-2it}} .$$

Designating $\exp it = z$ the reciprocal one-to-one mapping of the unit circle onto the single-contour Cassini's oval is obtained:

$$w = az \sqrt{1 + \frac{\lambda}{z^2}} , \tag{1.10.2}$$

where $\lambda = (c/a)^2 < 1$.

The function (1.10.2) has an analytic continuation into the exterior of the unit circle. It represents the single-valued branch of the two-valued function with branch points positioned inside the circle. The function (1.10.2) has the first order pole at the point at infinity. Therefore, in accordance with the boundary correspondence principle the function carries out the conformal mapping of the exterior of the unit circle onto the exterior of the single-contour Cassini's oval. When $a = c$ the function (1.10.2) maps the exterior of the unit circle onto the exterior of the Bernoulli's lemniscata.

The mapping of the polar net of lines in the exterior of the circle $|z| > 1$ carried out by the function (1.10.2) when $a = c = 1$ is presented in Fig. 1.47. Note that the circumferences $|z| = C$ $(C > 1)$ are mapped onto Cassini's

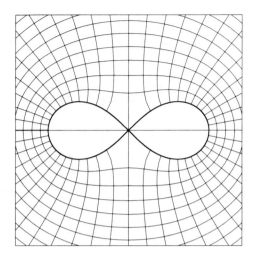

FIGURE 1.47

ovals and the rays $\arg z = $ const are mapped onto the arcs of equi-sided hyperbolas.

To map the finite domain bounded by one contour of the two-contour Cassini's oval (when $a < c$) or by one loop of Bernoulli's lemniscata (when $a = c = 1$) the parameter t is introduced in another way. Let

$$w^2 - c^2 = a^2\, e^{it},$$

whence

$$w = \sqrt{c^2 + a^2\, e^{it}}.$$

Assuming $e^{it} = z$ the function is obtained realizing the reciprocal one-to-one mapping of the unit circumference onto one contour of the two-contour Cassini's oval:

$$w = c\sqrt{1 + \lambda z},$$

where $\lambda = (a/c)^2 \leq 1$.

As this function is analytic inside the unit circle $|z| < 1$, it realizes the conformal mapping of the circle onto the domain bounded by one loop of the Cassini's oval or lemniscata.

1.10.3 The generalized Cassini's oval and the generalized lemniscata

The set of points in the w-plane satisfying the condition

$$|w^n - c^n| = a^n$$

is called the *generalized Cassini's oval* (or n-order cassinian).

Let us designate as w_k the roots of the equation $w^n - c^n = 0$, i.e.,

$$w_k = c\, e^{2\pi i k/n}\,,$$

where $k = 1, 2, \ldots, n$. Expanding the polynomial $w^n - c^n$ into the linear multipliers the equation of n-order cassinian can be written as

$$|w - w_1| \cdot |w - w_2| \cdots |w - w_n| = a^n\,.$$

Therefore, the n-order Cassini's oval is the set of points in the w-plane with the product of distances to the points w_1, $w_2, \ldots, w_n$ being constant and equal to a^n.

When $a > c$ the generalized Cassini's oval represents a smooth Jordan curve (the *single-contour generalized oval*). When $a = c$ the curve is a non-Jordan line with the self-intersection point at the origin (the *generalized lemniscata*). When $a < c$ the curve splits into n smooth loops not connected with each other (*n-contour Cassini's oval*). When $a \geq c$, i.e., in the case of the single-contour oval or lemniscata it is possible to parametrize the generalized Cassini's oval as

$$w^n = c^n + a^n\, e^{int}\,.$$

Designating $\exp(it) = z$ we obtain

$$w = az\left(1 + \frac{\lambda}{z^n}\right)^{1/n}\,, \tag{1.10.3}$$

where $\lambda = (c/a)^n \leq 1$.

This function maps a circumference onto a single-contour generalized Cassini's oval in a reciprocal one-to-one manner; it is analytic outside the unit circle in the domain $|z| > 1$ and has the first order pole at the point at infinity. Therefore, this function conformally maps (when $\lambda \leq 1$) the exterior of the unit circle onto the exterior of the single-contour generalized Cassini's oval (or the generalized lemniscata).

The mapping of the polar net of lines outside the circle $|z| > 1$ carried out by the function (1.10.3) when $\lambda = 1$, $n = 4$ is presented in Fig. 1.48. It should be mentioned that the circumferences $|x| = C$ $(C > 1)$ are mapped by the function (1.10.3) onto the generalized cassinians.

1.10.4   The Pascal's limaçon and the cardioid

The curve with the polar equation $\rho = 2a \cos \varphi + 2l$ or with parametric equation

$$w = 2(a \cos t + l)\, e^{it}\,, \tag{1.10.4}$$

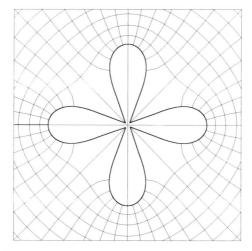

FIGURE 1.48

where $t = \varphi$, $-\pi \leq t \leq \pi$, $a > 0$, $l > 0$ is called the *Pascal's limaçon*. When $0 < l < a$ the curve is not the Jordan one, it has the self-intersection point at the origin if $\varphi = \pm \arccos(-l/a)$. Such a Pascal's limaçon is called the *hyperbolic* one. When $l = a$ the curve represents a cardioid that is the non-smooth Jordan line with the returning point[‡] at the origin. When $l > a$ the Pascal's limaçon is the smooth Jordan curve called the *elliptic limaçon*.

In the inversion of the Pascal's limaçon with respect to the unit circumference the curve with the polar equation

$$\rho = \frac{p}{1 + e \cos \varphi}$$

is obtained, where $p = 1/(2l)$, $e = a/l$. The last equation determines the conic curve with eccentricity equal to e and one of the foci placed at the origin. Therefore, the Pascal's limaçon represents the inversion of the conic curve with respect to the foci, namely, the elliptic (hyperbolic) limaçon is the inversion of an ellipse (hyperbola) and a cardioid is the inversion of a parabola.

Supposing $\exp(it) = z$ in the equation (1.10.4) we obtain the function which maps the circumference $|z| = 1$ onto the Pascal's limaçon

$$w = az^2 + a + 2lz = a\left(z^2 + 2\lambda z + 1\right), \qquad (1.10.5)$$

where $\lambda = l/a$. This function is analytic in the circle $|z| < 1$, when $l > a$

[‡]The *returning point* (or the cusp point) is such a singular point of a plane curve line in which two branches of the curve converging in this point have the common tangent and lie to the same side of the normal to the curve.

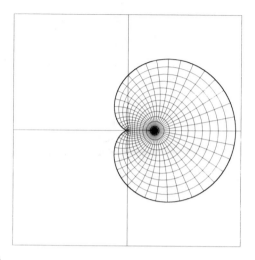

FIGURE 1.49

it reciprocally one-to-one maps the circumference $|z| = 1$ onto the Pascal's limaçon, therefore, in accordance with the boundary correspondence principle this function conformally maps the circle onto the finite domain bounded by the elliptic Pascal's limaçon or cardioid.

The mapping of the polar net in the unit circle carried out by the function (1.10.5) when $a = l = 1$ is represented in Fig. 1.49. It is remarkable, that the circumferences $|z| = C$ $(C < 1)$ are mapped onto the elliptic Pascal's limaçons and the radii of the circle ($\arg z = $ const) are mapped onto the parabola arcs.

The conformal mapping of the exterior of the Pascal limaçon onto the exterior of the circle is presented in Catalog 2 (Domain 20).

1.10.5 Generalized cardioids and lemniscatas

In the Subsection (1.7.7) it was established that the function

$$w = (1 + z)^\alpha \tag{1.10.6}$$

maps the circumference $z = e^{it}$ onto the curve with the polar equation

$$R = \left(2\cos\frac{\Phi}{\alpha}\right)^\alpha, \tag{1.10.7}$$

where $-\pi\alpha/2 < \Phi < \pi\alpha/2$. When $0 < \alpha < 2$ the curve (1.10.7) is the Jordan one with the angular point at the origin. In Subsections 1.10.2 and 1.10.4 the particular cases $\alpha = 1/2$ (one loop of Bernoulli's lemniscata) and $\alpha = 2$ (the cardioid) were considered.

In the case $0 < \alpha < 1$ the curve (1.10.7) is called the generalized lemniscata and when $1 < \alpha < 2$ — the generalized cardioid. When $0 < \alpha \leq 2$ the function (1.10.6) is univalent in the circle $|z| < 1$ and conformally maps it onto the domain bounded by the generalized lemniscata or cardioid (1.10.7).

When $|z| > 2$ the curve (1.10.7) is not a Jordan one and the function (1.10.6) is not univalent in the unit circle.

1.10.6 Epicycloids and shortened epitrochoids

A closed curve defined parametrically

$$u = a\left(n\cos t - \lambda \cos nt\right), \quad v = a\left(n\sin t - \lambda \sin nt\right)$$

or

$$w = a\left(n\,e^{it} - \lambda\,e^{int}\right), \tag{1.10.8}$$

where $-\pi \leq t \leq \pi$, n is a natural number greater than 1, is called *epitrochoid*. When $0 < \lambda < 1$ it represents a smooth Jordan curve (the *shortened epicycloid*). When $\lambda = 1$ it turns into the ordinary *epicycloid*, which is a non-smooth Jordan curve with $n - 1$ returning points. At last, when $\lambda > 1$ it represents the *oblong epicycloid* that is a non-Jordan curve with $n - 1$ self-intersection points. When $n = 2$ the epicycloid represents the cardioid and epitrochoid represents the Pascal's limaçon.

We will consider the mapping of the finite domain bounded by the Jordan epitrochoid ($\lambda \leq 1$). Assuming $\exp(it) = z$ in the formula (1.10.8) we obtain the function

$$w = a\left(nz - \lambda z^n\right), \tag{1.10.9}$$

which reciprocally one-to-one maps the unit circle $|z| < 1$ onto the epitrochoid. This function is analytic inside the circle $|z| < 1$, in accordance with the boundary correspondence principle it conformally maps the circle onto the finite domain bounded by the epitrochoid. The mapping of the polar net of lines carried out by the function (1.10.8) when $\lambda = 1$, $n = 4$ is depicted in Fig. 1.50. The boundary of the circle $|z| = 1$ is mapped onto the epicycloid and the circumferences $|z| = $ const are mapped onto the epitrochoids.

1.10.7 Hypocycloids and shortened hypotrochoids

The line defined parametrically

$$u = a\left(n\cos t + \lambda \cos nt\right), \quad v = a\left(n\sin t - \lambda \sin nt\right)$$

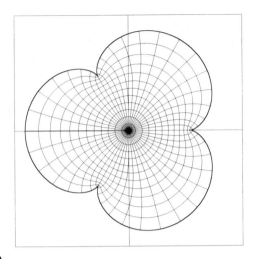

FIGURE 1.50

or

$$w = a \left(n\, e^{it} + \lambda\, e^{-int} \right), \qquad (1.10.10)$$

where $-\pi \le t \le \pi$, n is any natural number, is called the *hypotrochoid*. When $0 < \lambda < 1$ it is a smooth Jordan curve (the *shortened hypocycloid*), when $\lambda = 1$ it is a non-smooth Jordan curve with $n + 1$ returning points (the ordinary *hypocycloid*), when $\lambda > 1$ it is a non-Jordan curve with $n + 1$ self-intersecting points (the *oblong hypocycloid*). We will consider only the case of Jordan curves ($\lambda \le 1$).

Assuming $\exp(it) = z$ in formula (1.10.10) we obtain the function

$$w = a \left(nz + \frac{\lambda}{z^n} \right), \qquad (1.10.11)$$

which reciprocally one-to-one maps the circumference $|z| = 1$ onto the hypotrochoid. The function (1.10.11) is analytic in the exterior of the circle $|z| > 1$, it has the first order pole at the point at infinity. In accordance with the boundary correspondence principle this function carries out the conformal mapping of the exterior of the circle onto that of the hypocycloid.

The mapping of the polar net of lines carried out by the function (1.10.11) ($\lambda = 1$, $n = 3$) is presented in Fig. 1.51. The circumference $|z| = 1$ is mapped onto the *astroid* and the circumferences $|z| = r > 1$ are mapped onto the shortened hypocycloid.

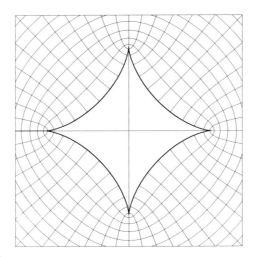

FIGURE 1.51

1.10.8 The exterior of a star-shaped cut

Let us consider the function

$$
w = \frac{z}{\sqrt[n]{4}} \left(1 + \frac{1}{z^n} \right)^{2/n}, \tag{1.10.12}
$$

where n is a natural number. When $n = 1$ and $n = 2$ this function is a rational one, when $n = 2$ it represents the Zhukovskii function and when $n = 1$ it is obtained from the Zhukovskii function with the help of the linear transformation:

$$
w = \frac{1}{4} \left(z + \frac{1}{z} \right) + \frac{1}{2}.
$$

The Zhukovskii function is univalent in the exterior of the unit circle $|z| > 1$ and conformally maps it onto the infinite w-plane with the cut along the interval $(-1, 1)$. It conformally maps the polar net of lines onto the orthogonal net of confocal ellipses and hyperbolas. Therefore, when $n = 2$ the function (1.10.12) maps the polar net in the z-plane onto the net of ellipses and hyperbolas with the center at the origin and foci at the points $w = -1$ and $w = 1$. When $n = 1$ the function (1.10.12) conformally maps the exterior of the circle $|z| > 1$ onto that of the interval $(0, 1)$ and the polar net in the z-plane onto the net of ellipses and hyperbolas with the foci at the points $w = 0$ and $w = 1$. When $n > 2$ the function (1.10.12) is a multiply-valued one with the branch points $z_k = \exp(\pi i(2k - 1)/n)$ lying on the circumference $|z| = 1$ (here $k = 1, 2, \dots, n$). Outside the circle $|z| > 1$ the function (1.10.12) splits into n single-valued branches. We will consider the principal branch selected by the condition $w(1) = 1$.

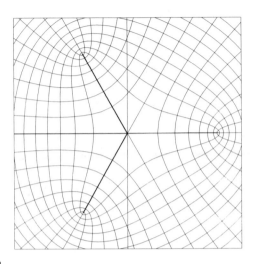

FIGURE 1.52

In the circumference $|z| = 1$ besides the branch points z_k we mark the points

$$z_k' = e^{2\pi i k/n}, \quad k = 1, 2, \ldots, n.$$

The function (1.10.12) maps the points z_k' into themselves and the points z_k into the origin. If $z = e^{i\varphi}$ we have

$$w = \left(\cos\frac{n\varphi}{2}\right)^{2/n},$$

where one of the values of the n-valued function is assumed. More precisely if the point $z = e^{i\varphi}$ lies in the arc $\pi(2k-1)/n < \varphi < \pi(2k+1)/n$ between two adjoining branch points, then

$$w = \left|\cos\frac{n\varphi}{2}\right|^{2/n} e^{2\pi i k/n}.$$

Therefore, the arc of the unit circumference between two branch points is mapped by the function (1.10.12) onto a twice traversed segment, which connects the points 0 and $z_k' = \exp(2\pi i k/n)$. All the circumference $|z| = 1$ is mapped onto a star-shaped cut formed by n symmetrical segments emerging from the origin. The function (1.10.12) is analytic outside the circle $|z| > 1$ and it has the first order pole at the point at infinity, therefore in accordance with the boundary correspondence principle it conformally maps the exterior of the unit circle onto the exterior of the star-shaped cut.

The mapping of the polar net of lines carried out by the function (1.10.12) when $n = 3$ is presented in Fig. 1.52. Note, that circumferences $|z| = r > 1$ are mapped onto the smooth Jordan curves that can be called the *smoothed n-pointed stars.* The conformal mapping of the exterior of the unit circle onto the exterior of such a smoothed star is carried out by the function

$$w = Cz \left(1 + \frac{\lambda}{z^n} \right)^{2/n} ,$$

where $0 < \lambda < 1$.

1.10.9 The plane with n symmetrical cuts along the rays. Booth's lemniscatas

Having made the inversion transformation in the function (1.10.12) both for the z and w-variables we obtain

$$w = z \left(\frac{2}{1 + z^n} \right)^{2/n} . \tag{1.10.13}$$

As well as in the previous Subsection, when $n > 2$ we consider the single-valued branch of the multiply-valued function, selected by the condition $w(1) = 1$. It is analytic in the circle $|z| < 1$ and maps the boundary of the circle onto n infinite cuts along the symmetrically located rays

$$w = R e^{2\pi i k/n} ,$$

where $k = 1, 2, \ldots, n$, $1 < R < \infty$. In accordance with the boundary correspondence principle the function (1.10.13) conformally maps the unit circle onto the infinite plane with cuts along the rays mentioned above.

The cases $n = 1$ and $n = 2$ should be considered particularly. When $n = 1$ the function (1.10.13) can be represented as a sequence of two transformations

$$t = \frac{1}{4} \left(z + \frac{1}{z} \right) + \frac{1}{2} , \quad w = \frac{1}{t} .$$

As was mentioned in the previous Subsection the first transformation maps the polar net in the z-plane onto the net of ellipses and hyperbolas with foci at the points $t = 0$ and $t = 1$. In the second transformation (the inversion with respect to the point $t = 0$ being one of the foci) the conic curves are mapped onto the Pascal's limaçons (Subsection 1.10.4). Therefore, the function

$$w = 4 \frac{z}{(z + 1)^2} \tag{1.10.14}$$

conformally maps the circle $|z| < 1$ onto the w-plane with the cut along the ray of the real axis $(1, \infty)$, the circumferences $|z| = $ const being mapped onto the elliptic Pascal's limaçons and radii $\arg z = $ const being mapped onto the hyperbolic Pascal's limaçons. The mapping of the circle carried out by the function (1.10.14) is presented in Fig. 1.53.

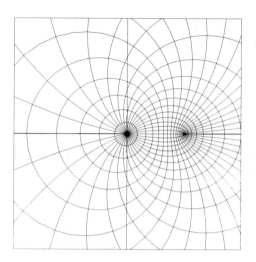

FIGURE 1.53

When $n = 2$ the function (1.10.13) can be represented as a sequence of transformations

$$t = \frac{1}{2}\left(z + \frac{1}{z}\right), \quad w = \frac{1}{t}.$$

The first transformation (the Zhukovskii function) conformally maps the unit circle onto the t-plane with the cut along the interval $(-1, 1)$ and the polar net — onto the net of ellipses and hyperbolas. The second transformation (the inversion with respect to the center of the conic curves) maps ellipses and hyperbolas onto the curves called the Booth's lemniscatas (see Subsection 1.7.8).

The net of the reciprocally-orthogonal Booth's lemniscatas obtained in the mapping of the polar net in the unit circle $|z| < 1$ by the function

$$w = 2\,\frac{z}{z^2 + 1}$$

is presented in Fig. 1.54. The depicted net of lines differs from that in Fig. 1.34 only in the density of lines near the origin.

In the general case the function (1.10.13) maps the polar net in the circle $|z| < 1$ onto the orthogonal net in the w-plane with n cuts along the symmetrically positioned rays. The lines of the net can be called the *generalized Booth's lemniscatas*. The mapping of the polar net in the unit circle realized by the function (1.10.13) when $n = 4$ is presented in Fig. 1.55.

More general function

$$w = Cz\,(1 + \lambda z^n)^{-2/n}$$

with $0 < \lambda < 1$ realizes the conformal mapping of the unit circle onto

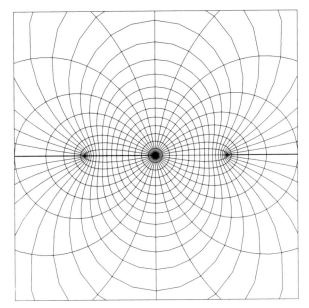

FIGURE 1.54

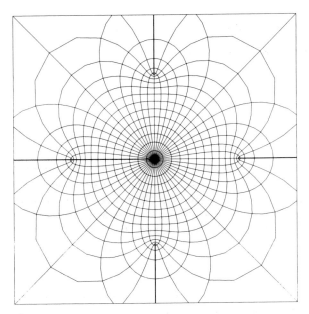

FIGURE 1.55

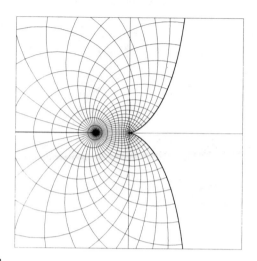

FIGURE 1.56

the interior of the smooth Jordan curve, which is the generalized Booth's lemniscata.

1.10.10   The cissoid

The curve with the polar equation

$$\rho = \sec \varphi - \cos \varphi$$

or

$$w = \rho\, e^{i\varphi} = \frac{2e^{2i\varphi}}{e^{2i\varphi}+1} - \frac{e^{2i\varphi}+1}{2},$$

where $-\pi/2 < \varphi < \pi/2$ is called the *cissoid*. It represents an infinite curve without self-intersections with the returning point $w = 0$ (as $\varphi = 0$). We consider it as a Jordan curve of the extended complex plane. Assuming $\exp(2i\varphi) = z$ $(-\pi < \arg z < \pi)$ we obtain the reciprocal one-to-one mapping of the unit circumference onto the cissoid:

$$w = -\frac{1}{2}\frac{(z-1)^2}{z+1}. \tag{1.10.15}$$

The function (1.10.15) is analytic in the circle $|z| < 1$, in accordance with the boundary correspondence principle it conformally maps the unit circle $|z| < 1$ onto the infinite domain bounded by the cissoid.

The mapping of the polar net of lines in the unit circle carried out by this function is presented in Fig. 1.56.

Note that the function (1.10.15) is expressed through the Zhukovskii function. Indeed, introducing a new variable $t = (z+1)/2$ we obtain

$$w = 2 - \left(t + \frac{1}{t}\right).$$

Therefore, the cissoid is the image of the circumference $|2t - 1| = 1$ in the mapping carried out by the Zhukovskii function.

1.10.11 The strophoid

The infinite curve defined by the polar equation

$$\rho = \sec\varphi - 2\cos\varphi$$

or

$$w = \rho\, e^{i\varphi} = \frac{2e^{2i\varphi}}{e^{2i\varphi} + 1} - e^{2i\varphi} - 1, \tag{1.10.16}$$

where $-\pi/2 < \varphi < \pi/2$ is called the *strophoid*. The strophoid is not a Jordan curve, it has the self-intersection point at the origin (as $\varphi = \pm\pi/4$).

The algebraic parametrization of the strophoid is obtained by introducing a parameter $t = \tan\varphi$ into the equation (1.10.16):

$$w = (1 - 2\cos^2\varphi)(1 + i\tan\varphi) = \frac{t^2 - 1}{1 - it}. \tag{1.10.17}$$

Designating $\exp(2i\varphi)$ as z in the formula (1.10.16) we obtain the function, which maps the unit circumference $|z| = 1$ onto the strophoid:[§]

$$w = \frac{2z}{z+1} - z - 1 = -\frac{z^2 + 1}{z + 1}. \tag{1.10.18}$$

The function (1.10.18) is analytic in the circle $|z| < 1$ but is not univalent. Indeed, its derivative

$$w' = -1 + \frac{2}{(z+1)^2}$$

vanishes at points $z = -1 \pm \sqrt{2}$, one of which lies inside the circle. Nevertheless in the half-circle $|z| < 1$, $\operatorname{Re} z < 0$ the function (1.10.18) is univalent. It is notable that the vertical diameter of the circle (the interval $(-i, i)$) is mapped onto the loop of the strophoid. Indeed, assuming $z = i\tau$ in the formula (1.10.18) we obtain

$$w = \frac{\tau^2 - 1}{1 - i\tau},$$

[§] Note that the function (1.10.18) is reduced to the Zhukovskii function by the linear transformation. Therefore, the strophoid is the image of some circumference in the mapping carried out by the Zhukovskii function.

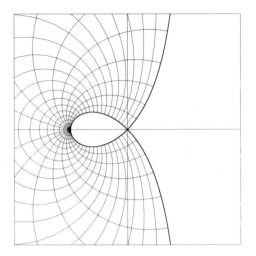

FIGURE 1.57

this precisely agrees with (1.10.17). Therefore, the half-circle $|z| < 1$, Re $z < 0$ is conformally mapped by the function (1.10.18) onto the singly connected domain bounded by the strophoid. The mapping of the polar net of lines in the half-circle is presented in Fig. 1.57.

The function (1.10.18) maps the right half-plane with the half-circle deleted (i.e., the domain Re $z > 0$, $|z| > 1$) onto exactly the same domain.

1.10.12   The tractrix

The curve defined by the parametric equation

$$u = \ln\left(\tan\frac{t}{2}\right) + \cos t, \quad v = \sin t$$

or

$$w = \ln\left(\tan\frac{t}{2}\right) + e^{it} \quad (0 < t < \pi)$$

is called the *tractrix*. It represents an infinite curve with the returning point $t = \pi/2$. It is more convenient to consider the curve shifted by the distance $-\pi/2$ along the imaginary axis:

$$w = \ln\left(\tan\frac{t}{2}\right) + e^{it} - i\,\frac{\pi}{2}. \tag{1.10.19}$$

Assuming $\xi = -\exp(it)$ and expressing $\tan(t/2)$ through ξ by the formula

$$\tan \frac{t}{2} = i \frac{1 - e^{it}}{1 + e^{it}} = i \frac{1 + \xi}{1 - \xi}$$

we obtain the equation of the tractrix (1.10.19) in the form

$$w = \ln \frac{1 + \xi}{1 - \xi} - \xi . \tag{1.10.20}$$

The function (1.10.20) maps the lower half-circumference $|\xi| = 1$, $-\pi < \arg \xi < 0$ onto the tractrix (1.10.19). It is real-valued if the parameter ξ is real. Due to the symmetry principle the function (1.10.20) maps the upper half-circumference $0 < \arg \xi < \pi$ onto the curve symmetrical to the tractrix (1.10.19) with respect to the real axis. Therefore, the function (1.10.20) analytic in the circle $|\xi| < 1$ maps the circumference $|\xi| = 1$ reciprocally one-to-one onto the boundary of the infinite domain between two symmetrical tractrices. In accordance with the boundary correspondence principle the circle $|\xi| < 1$ is conformally mapped onto the mentioned curvilinear strip enclosed between two tractrices.

Making the change of the independent variable

$$\xi = \tanh \frac{z}{2}$$

we obtain the function realizing the mapping of the strip $-\pi/2 < \operatorname{Im} z < \pi/2$ onto the curvilinear strip enclosed between two tractrices:

$$w = z - \tanh \frac{z}{2} . \tag{1.10.21}$$

The mapping of the Cartesian net of lines in the strip $-\pi/2 < \operatorname{Im} z < \pi/2$ onto the isothermic net of lines in the curvilinear strip is presented in Fig. 1.58.

Exercises

1. With the help of the program CONFORM create the mapping of the exterior of the circle $|z| > 1$ realized by the function

$$w = z \left(1 + \frac{1}{z^n} \right)^p , \tag{1.10.22}$$

where n is a natural number, $p > 0$ is a real number. Show that if $0 < p < 2/n$ the exterior of the circle is conformally mapped onto the exterior of the symmetrical n-sheet figure and if $p > 2/n$ the mapping is not conformal.

Remark. The particular cases of the function (1.10.22) were considered above: namely, when $p = 2/n$ the exterior of the circle is mapped onto the exterior of the star-shaped cut (Subsection 1.10.8), and when $p =$

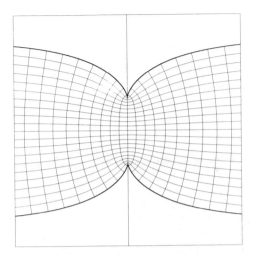

FIGURE 1.58

$1/n$ it is mapped onto the exterior of the lemniscata of the n-th order (Subsection 1.10.3).

2. The function

$$w = \frac{z}{(1 + z^n)^p} \tag{1.10.23}$$

is obtained from (1.10.22) by inversion with respect to both z and w. With the help of the program CONFORM, show that if $0 < p < 2/n$ the unit circle $|z| < 1$ is conformally mapped by the function (1.10.23) onto the infinite domain (the strip with n branches), and if $p > 2/n$ the function (1.10.23) is not univalent in the circle $|z| < 1$.

Remark. The particular case of the function (1.10.23) when $p = 2/n$ is considered above in the Subsection 1.10.9.

3. Create the mapping of the exterior of the unit circle $|z| > 1$ realized by the function

$$w = z \left[\left(1 + \frac{1}{z} \right) \left(1 + \frac{1}{z^2} \right) \right]^p . \tag{1.10.24}$$

Show that if $0 < p \le 2/3$ the exterior of the circle is conformally mapped by the function (1.10.24) onto the exterior of the non-symmetrical three-sheet figure, and when $p > 2/3$ the mapping is not conformal.

1.11 The Construction of Conformal Mappings of Curvilinear Strips with the Help of an Analytic Continuation of a Function from the Real Axis

The problems on conformal mappings considered in the previous Section represent examples of an analytic continuation of functions given on a circumference inside or outside the circumference. In this Section problems on analytic continuations of functions given on an infinite straight line are considered.

Let L be an infinite curve without self-intersections that is given parametrically

$$u = u(t), \quad v = v(t)$$

or

$$w = u(t) + iv(t),$$

where $u(t)$, $v(t)$ are real-valued analytic functions of the real variable t ($t \in \mathcal{R}$). Assume $|w(t)| \to \infty$ as $|t| \to \infty$. We consider such a curve as an infinite Jordan line on the extended complex plane w. Suppose for simplicity that the curve L is a smooth one.

The function $w(t)$ primarily given on the real axis t has the analytic continuation $w(z)$ into some strip of the complex plane $-\alpha < \operatorname{Im} z < \beta$, where $\alpha > 0$, $\beta > 0$. If the function $w(z)$ turns out to be univalent in the strip, it conformally maps the strip onto some neighborhood of the curve L, which represents a curvilinear strip of the w-plane. In particular, the strip $0 < \operatorname{Im} z < \beta$ (or $-\alpha < \operatorname{Im} z < 0$) is mapped onto the curvilinear strip bounded by the curve L on one of the sides.

The function $w(z)$ can happen to be analytic and univalent in the upper half-plane $\operatorname{Im} z > 0$ (or in the lower half-plane $\operatorname{Im} z < 0$). In this case it realizes the conformal mapping of the upper (lower) half-plane onto the curvilinear angular domain positioned to the left (to the right) from the curve L.*

Let us consider conformal mappings of some curvilinear strips with the help of functions obtained by the analytic continuation of the function $w(t)$ from the real axis.

First note that to continue the function $w(t)$ analytically into the strip $0 < \operatorname{Im} z < \beta$ (or into the strip $-\alpha < \operatorname{Im} z < 0$) the demand of smoothness of the curve L is not necessary and it is not valid for all considered examples.

1.11.1 The parabola

Let the parabola be given by the equation

$$v^2 = 2pu$$

or

$$w(t) = \frac{t^2}{2p} + it,$$

---

*The traversal sense of the curve L corresponds to the increasing of the parameter t.

where $p > 0$, $t \in \mathcal{R}$. The function $w(t)$ has the analytic continuation from the real axis onto the whole complex plane:

$$w(z) = \frac{z^2}{2p} + iz = \frac{(z + ip)^2}{2p} + \frac{p}{2}, \quad z \in \mathcal{C}. \qquad (1.11.1)$$

The function (1.11.1) satisfies the necessary condition of the univalence at all points of the z-plane, where

$$w' = \frac{z + ip}{p} \neq 0.$$

Therefore, the function (1.11.1) is locally univalent everywhere except for the point $z = -ip$.

The function (1.11.1) is univalent in the half-plane Im $z > 0$ and conformally maps it onto the exterior of the considered parabola.[†] In the strip $-p < $ Im $z < p$ the function (1.11.1) is also univalent and conformally maps the strip onto the interior of the parabola with the cut along the ray of the real axis $(p/2, \infty)$.

The conformal mapping of the interior of the parabola (without cuts) is presented in Catalog 3 (Domain 45).

1.11.2   The hyperbola

The parametric equation of the branch of the hyperbola has the form

$$u = a \cosh t, \quad v = b \sinh t$$

or

$$w(t) = a \cosh t + ib \sinh t,$$

where $a > 0$, $b > 0$, $t \in \mathcal{R}$.

Let us consider the analytic continuation of the function $w(t)$ from the real axis into the whole complex z-plane:

$$w(z) = a \cosh z + ib \sinh z, \quad z \in \mathcal{C}. \qquad (1.11.2)$$

The function (1.11.2) is locally univalent at all points of the z-plane where

$$w' = a \sinh z + ib \cosh z \neq 0.$$

The necessary condition of the univalence is violated at the points where

$$\tanh z = \frac{ib}{a},$$

[†]An infinite singly connected domain bounded by the parabola and not containing the focus of the parabola is called its exterior. The complement of this domain is called the interior of the parabola.

or

$$z = i\left(-\arctan\frac{b}{a} + \pi n\right).$$

The function (1.11.2) is univalent in the strip $0 < \operatorname{Im} z < \pi - \arctan(b/a)$ and conformally maps it onto the infinite curvilinear strip bounded by the considered hyperbola and the cut along the ray of the real axis $(-\infty, -\sqrt{a^2 + b^2})$. In the similar way the strip $-\arctan(b/a) < \operatorname{Im} z < 0$ is conformally mapped onto the curvilinear strip between the considered hyperbola and the cut along the ray of the real axis $(\sqrt{a^2 + b^2}, \infty)$.

The conformal mapping of the half-plane onto the exterior of one branch of the hyperbola (without cuts) is presented in Catalog 3 (Domain 54).

1.11.3 The cissoid and ellipses inversions

Let us consider infinite curves determined by the polar equation

$$\rho = C\left(\sec\varphi - \lambda\cos\varphi\right),$$

or

$$w = C\left(1 - \lambda\cos^2\varphi\right)\left(1 + i\tan\varphi\right), \tag{1.11.3}$$

where $-\pi/2 < \varphi < \pi/2$, $C > 0$, $\lambda > 0$. The curves (1.11.3) are considered as closed lines of the extended complex plane, that pass (once) through the point at infinity. In the previous Section particular cases of the curve (1.11.3) were considered: when $\lambda = 1$ the line represents the cissoid (Subsection 1.10.10) and when $\lambda = 2$ it is the strophoid (Subsection 1.10.11).

When $0 < \lambda < 1$ the curves (1.11.3) are smooth Jordan lines. When $\lambda > 1$ curves (1.11.3) are not Jordan ones, they have the self-intersection point at the origin (if $\varphi = \pm\arccos(1/\sqrt{\lambda})$).

The algebraic parametrization of the curves (1.11.3) can be obtained by introducing the parameter $t = \tan\varphi$:

$$w = C + C\left(it - \frac{\lambda}{1 - it}\right) \tag{1.11.4}$$

or

$$u = C\left(1 - \frac{\lambda}{1 + t^2}\right), \quad v = Ct\left(1 - \frac{\lambda}{1 + t^2}\right).$$

Finally, excluding the parameter t from the last equations we obtain the implicit equation of the curves (1.11.3)

$$u = C\left(1 - \frac{\lambda u^2}{u^2 + v^2}\right). \tag{1.11.5}$$

Carrying out transformation of the inversion with the curves (1.11.5), i.e., substituting u and v for $u/(u^2+v^2)$ and $v/(u^2+v^2)$ respectively, we obtain the equation of conic curve

$$\frac{u}{C} = (1-\lambda)\, u^2 + v^2\,,$$

which determines ellipse, hyperbola or parabola passing through the origin depending on the sign of $1-\lambda$. Therefore, when $\lambda = 1$ the curve (1.11.3) (the cissoid) represents the inversion of the parabola with respect to its vertex and when $\lambda < 1$ ($\lambda > 1$) it represents the inversion of the ellipse (hyperbola) with respect to one of the vertexes of the ellipse (hyperbola).

The curves (1.11.3) can be considered as images of vertical straight lines $\operatorname{Re} z = -C$ in the mapping carried out by the Zhukovskii function. Indeed, the function $w_1 = z + 1/z$ maps straight lines $z = -C + iCt = -C\,(1-it)$ into

$$w_1 = -C + C\left(it - \frac{1}{C^2(1-it)}\right).\tag{1.11.6}$$

The function (1.11.4) differs from (1.11.6) only by the sign of the first (constant) term, $\lambda = 1/C^2$ being substituted into (1.11.4), therefore

$$w = z + \frac{1}{z} + 2C\,,$$

where $z = -C\,(1-it)$.

Assuming $C = 1$ ($\lambda = 1$) we obtain the function

$$w = z + \frac{1}{z} + 2\,,\tag{1.11.7}$$

which maps the straight line $z = -1 + it$ onto the cissoid. The analytic continuation of the function (1.11.7) represents the univalent function in the half-plane $\operatorname{Re} z < -1$. It conformally maps the half-plane onto the infinite domain bounded by the cissoid. The mapping of the Cartesian net of lines in the half-plane $\operatorname{Re} z < -1$ carried out by the function (1.11.7) is presented in Fig. 1.59. Note that images of straight lines $\operatorname{Re} z = \mathrm{const}$ are inversions of ellipses that differ from the curves (1.11.4) only by the position.

1.11.4   The strophoid and other inversions of the hyperbola

Let us consider the curves (1.11.3) when $\lambda > 1$. As was shown in the previous Subsection, when $\lambda > 1$ the curve represents the inversion of the hyperbola with respect to one of its vertexes. The curve (1.11.3) has a loop and splits the w-plane into three singly connected domains. Let us consider the conformal mapping of one of them bounded by the complete curve (1.11.3) — the loop and the infinite branches — carried out onto

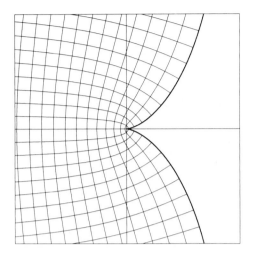

FIGURE 1.59

the half-plane. Denote this domain as G. Consider the curve (1.11.3) as the image of a vertical straight line Re $\xi = -C$ (where $0 < C < 1$, $\lambda = 1/C^2 > 1$) in the mapping carried out by the Zhukovskii function

$$w = \xi + \frac{1}{\xi} + 2C\,.$$

It follows from the properties of the Zhukovskii function (Subsection 1.7.4), that the straight line Re $\xi = -C$ does not lie in the domain of univalence of the Zhukovskii function when $0 < C < 1$, therefore the mapping of any strip containing this straight line can not be conformal. Nevertheless it is possible to specify such a domain of the plane in which the Zhukovskii function is univalent and the boundary of which is mapped onto the curve (1.11.3). Indeed, the circumference

$$\left| \xi + \frac{1}{2C} \right| = \frac{1}{2C}\,,$$

passing through the origin is symmetrical to the straight line Re $\xi = -C$ with respect to the circumference $|\xi| = 1$. Due to the Zhukovskii function property the noted circumference and the straight line are mapped onto the same line (1.11.3). In the domain Re $\xi < -C$, $|\xi + 1/(2C)| > 1/(2C)$ (the circular lune) the Zhukovskii function is univalent, therefore, the domain is conformally mapped by the Zhukovskii function onto the considered domain G.

To construct the conformal mapping of the domain G onto the half-plane

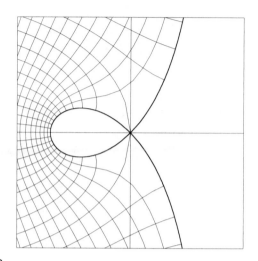

FIGURE 1.60

it is necessary to map the circular lune (two-sided polygon)

$$\operatorname{Re} \xi < -C, \quad \left| \xi + \frac{1}{2C} \right| > \frac{1}{2C}$$

onto it. Designate $C = \cos p$ $(0 < p < \pi/2)$ and use formulas of Subsection 1.9.1. The mapping of the considered circular lune onto the half-plane $\operatorname{Im} z > 0$ is realized by the sequence of transformations:

$$s = -i \frac{\xi + \cos p}{\sin p}, \quad r = \frac{s - 1}{s + 1},$$

$$u = r^{\pi/(\pi - 2p)}, \quad z = \frac{1 + u}{1 - u}.$$

The inverse mapping of the half-plane $\operatorname{Im} z > 0$ onto the domain G is carried out by the sequence of transformations

$$u = \frac{z - 1}{z + 1}, \quad r = u^{1 - 2p/\pi}, \quad s = \frac{1 + r}{1 - r},$$

$$\xi = is \sin p - \cos p, \quad w = \xi + \frac{1}{\xi} + 2 \cos p. \tag{1.11.8}$$

The mapping of the Cartesian net of lines carried out by the function (1.11.8) when $p = \pi/4 \approx 0.785$ is presented in Fig. 1.60. In this case the boundary of the curvilinear half-plane is the strophoid.

1.11.5 The semicubical parabola

The curve defined by the equation $v = u^{2/3}$ is called the *semicubical parabola* (or Neil's parabola). It is more convenient to consider the curve shifted by the distance $2/27$ along the ordinate axis:

$$v = u^{2/3} + \frac{2}{27}. \tag{1.11.9}$$

The curve (1.11.9) can be represented in the parametric form:

$$u = -t^3, \quad v = t^2 + \frac{2}{27}$$

or

$$w = -t^3 + it^2 + i\frac{2}{27},$$

where $t \in \mathcal{R}$. Introducing the complex variable $z = t + i/3$ we obtain

$$w(z) = -z^3 - \frac{z}{3}. \tag{1.11.10}$$

The function (1.11.10) maps the straight line $z = t + i/3$ onto the Neil's parabola (1.11.9), it has the analytic continuation onto the whole complex z-plane and takes real values for real values of z. In accordance with the symmetry principle the function (1.11.10) maps the symmetrical (with respect to the real axis) straight line $z = t - i/3$ onto the curve symmetrical to the curve (1.11.9) with respect to the real axis:

$$v = -u^{2/3} - \frac{2}{27}.$$

The function (1.11.10) satisfies the necessary condition of the univalence everywhere except for the points $z = \pm i/3$. In the strip $-1/3 < \operatorname{Im} z < 1/3$ the function (1.11.10) is univalent. It conformally maps the strip onto the curvilinear strip

$$|v| < u^{2/3} + \frac{2}{27},$$

bounded between two symmetrical Neil's parabolas.

The mapping of the Cartesian net in the strip $-1/3 < \operatorname{Im} z < 1/3$ realized by the function (1.11.10) is presented in Fig. 1.61.

1.11.6 The catenary

The curve defined by the equation

$$v = \cosh u$$

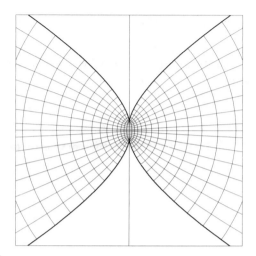

FIGURE 1.61

is called the *catenary*. It is more convenient to consider the curve

$$v = \cosh u + \frac{\pi}{2}\,, \tag{1.11.11}$$

which represents the catenary shifted by the distance $\pi/2$ along the imaginary axis. Its parametric equation has the form

$$w(t) = t + i \cosh t + i\frac{\pi}{2}\,,$$

where $t \in \mathcal{R}$.

Introducing a new variable $z = t + i\pi/2$ we obtain

$$w = z + \sinh z\,. \tag{1.11.12}$$

The function (1.11.12) is analytic in the whole complex plane. It maps the horizontal straight line $z = t + i\pi/2$ onto the catenary (1.11.11). Since this function is real-valued for real values of z, in accordance with the symmetry principle it maps the symmetrical (with respect to the real axis) straight line $z = t - i\pi/2$ onto another catenary

$$v = -\cosh u - \pi/2\,,$$

symmetrical to the curve (1.11.11) with respect to the real axis. The function (1.11.12) is univalent in the strip $-\pi/2 < \operatorname{Im} z < \pi/2$ and conformally maps the strip onto the infinite domain enclosed between two catenaries. Such mapping is presented in Fig. 1.62.

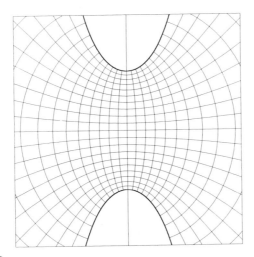

FIGURE 1.62

1.11.7   The curve $v = \sinh au$

The curve defined by the equation

$$v = \sinh au$$

can be represented in the parametric form if to designate $a = 1/p$ and to introduce the parameter $t = au = u/p$:

$$w(t) = pt + i \sinh t, \quad t \in \mathcal{R}.$$

This function defined primarily on the real axis has the analytic continuation into the whole complex plane:

$$w(z) = pz + i \sinh z, \quad z \in \mathcal{C}. \tag{1.11.13}$$

The function (1.11.13) is locally univalent at all points of the z-plane, where

$$w' = p + i \cosh z \neq 0.$$

The necessary condition of univalence is violated at points

$$z = \pm \left[\ln \left(p + \sqrt{p^2 + 1} \right) + i \frac{\pi}{2} \right] + 2\pi i n.$$

In the strip $-\pi/2 < \operatorname{Im} z < \pi/2$ the function (1.11.13) is univalent. The upper boundary of the strip, i.e., the straight line $z = t + i\pi/2$ is mapped by the function (1.11.13) onto the line

$$w = pt - \cosh t + \frac{i\pi p}{2}, \tag{1.11.14}$$

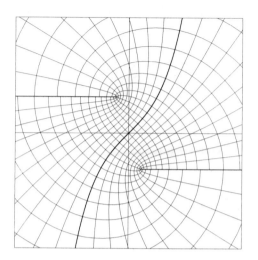

FIGURE 1.63

which represents the straight line cut — the ray

$$\text{Im } z = \frac{\pi p}{2}, \quad \text{Re } w < p \ln\left(p + \sqrt{p^2 + 1}\right) - \sqrt{p^2 + 1},$$

passed twice.

The function (1.11.13) is odd, so the straight line $z = t - i\pi/2$ is mapped onto the cut along a ray, which is symmetrical to the cut (1.11.14) with respect to the origin. All the strip $-\pi/2 < \text{Im } z < \pi/2$ is conformally mapped onto the w-plane with two cuts along oppositely directed rays of collinear straight lines.

The mapping of the Cartesian net in the strip $-\pi/2 < \text{Im } z < \pi/2$ realized by the function (1.11.13) with $p = 1$ is presented in Fig. 1.63. The curve $v = \sinh u$ is shown as a bold line.

Another solution of the problem on the mapping of the strip onto the plane with two infinite oppositely directed cuts along rays of collinear straight lines will be given in Subsection 1.12.2.

1.11.8   The exponential curve and Kirchhoff's curves

Let us consider the curve

$$v = e^u + \frac{\pi}{2} \tag{1.11.15}$$

in the plane $w = u + iv$, which represents an ordinary exponential curve shifted along the imaginary axis. The parametric equation of the curve (1.11.15) can be written as

$$w = t + i e^t + \frac{\pi i}{2}, \quad t \in \mathcal{R}.$$

Introducing a new variable $z = t + \pi i/2$ we obtain

$$w = z + e^z. \tag{1.11.16}$$

The function (1.11.16) is analytic in the whole complex plane. It maps the straight line $z = t + i\pi/2$ onto the curve (1.11.15) and the real axis onto itself. In accordance with the symmetry principle the function maps symmetrical (with respect to the real axis) straight line $z = t - i\pi/2$ onto the line $v = -e^u - \pi/2$ symmetrical to the curve (1.11.15) with respect to the real axis.

In the strip $-\pi/2 < \mathrm{Im}\, z < \pi/2$ the function is univalent and conformally maps the strip onto the domain enclosed between two exponential curves:

$$|v| < e^u + \pi/2. \tag{1.11.17}$$

Broader domain of univalence of the function (1.11.16) can be shown. The mapping realized by it is not locally conformal at points where

$$w' = 1 + e^z = 0,$$

i.e., at points $z = i\pi(2n + 1)$. For example, the function (1.11.16) is univalent in the strip $-\pi < \mathrm{Im}\, z < \pi$, which does not contain the noted points. The boundary line of the strip $z = t + i\pi$ is mapped by the function onto the curve

$$w = t - e^t + \pi i.$$

The imaginary part of this function is constant and the real part increases from $-\infty$ to -1 (when t varies from $-\infty$ to 0) and decreases from -1 to $-\infty$ with the further growth of t. Therefore, the straight line $z = t + i\pi$ is mapped onto a cut (twice passed ray $\mathrm{Im}\, w = \pi$, $\mathrm{Re}\, z \le -1$). The symmetrical straight line $z = t - i\pi$ is mapped by the function (1.11.16) onto the cut $\mathrm{Im}\, w = -\pi$, $\mathrm{Re}\, w < -1$. All the strip $-\pi < \mathrm{Im}\, z < \pi$ is conformally mapped by the function (1.11.16) onto the infinite w-plane with the cuts along rays of collinear straight lines. The mapping of the Cartesian net of lines in the strip is presented in Fig. 1.64.

The function (1.11.16) was found for the first time by G. R. Kirchhoff when the problem on the electric field in a capacitor formed by two parallel half-planes (see Subsection 2.2.4 for details) was solved. The net of lines in Fig. 1.64 represents the net of force lines and equipotentials in such a capacitor.

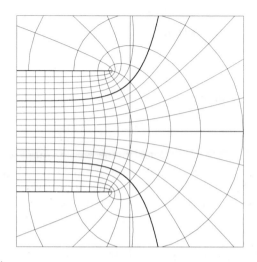

FIGURE 1.64

In particular the straight lines $z = t + iC$ are mapped onto equipotential curves with the parametric equation

$$w = t + iC + e^{t+iC}.$$

These lines can be called *Kirchhoff's curves*. The exponential curves

$$v = \pm(e^u + \pi/2) \tag{1.11.18}$$

represent two Kirchhoff's curves obtained when $C = \pm\pi/2$. They are depicted by bold lines in Fig. 1.64.

1.11.9   The cycloid and trochoids

The infinite curve with the parametric equation

$$u = t - \lambda \sin t, \quad v = \lambda \cos t - 1$$

or

$$w(t) = t + i\lambda\, e^{it} - i, \tag{1.11.19}$$

where $\lambda > 0$, $t \in \mathcal{R}$ is called the *trochoid*.

When $0 < \lambda < 1$ it represents a smooth infinite curve called the *shortened cycloid*. When $\lambda = 1$ the curve is an ordinary *cycloid*, which is a non-smooth curve with returning points $z = 2\pi n$. At last, when $\lambda > 1$ the curve is a line with self-intersections called the *oblong cycloid*.

When $\lambda \le 1$ the function (1.11.19) maps the real axis reciprocally one-to-one into the cycloid or the trochoid (the shortened cycloid). The function has the analytic continuation into all the complex plane:

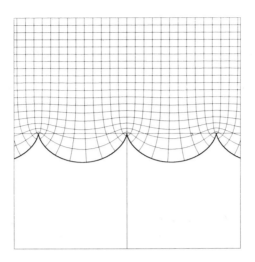

FIGURE 1.65

$$w(z) = z + i\lambda e^{iz} - i, \quad z \in \mathcal{C}. \tag{1.11.20}$$

The continued function has an essential singularity at the point at infinity, so the boundary correspondence principle is not applicable in the general form.

The derivative of the function (1.11.20)

$$w' = 1 - \lambda e^{iz}$$

when $\lambda \leq 1$ does not vanish at any point of the upper half-plane, i.e., the mapping realized by the function is locally conformal in the half-plane $\operatorname{Im} z > 0$.

The function (1.11.20) is univalent (when $\lambda \leq 1$) in the upper half-plane. It realizes the conformal mapping of the half-plane $\operatorname{Im} z > 0$ onto the infinite domain bounded by the cycloid or trochoid. The mapping of the Cartesian net of lines carried out by this function with $\lambda = 1$ is represented in Fig. 1.65. This isothermic net represents the net of equipotentials and streamlines for the flow of an ideal liquid in the curvilinear half-plane bounded by the cycloid (see Section 2.11 for details). Note that all the streamlines represent shortened trochoids.

Another domain of univalence of the function (1.11.20) can be found, namely, the strip $0 < \operatorname{Re} z < 2\pi$. The function (1.11.20) conformally maps this domain onto the infinite w-plane with two cuts along rays of parallel straight lines, i.e., onto the domain considered in the previous Subsection.

The function (1.11.20) is reduced to (1.11.16) by the linear transformations of variables z and w. Therefore, the problem on the potential

streaming around the infinite cycloid and the Kirchhoff's problem on the
electric field near the edge of a flat capacitor are equivalent. It follows that
equipotentials of the velocity field in the problem on streaming around the
cycloid are Kirchhoff's lines and each force line in the Kirchhoff's problem
represents the arch of a trochoid, which is the shortened cycloid inside the
capacitor and the oblong cycloid outside it.

Exercises

1. *Cubic parabola.* Represent the curve $v = u^3$ in the parametric form $w(t) = t + it^3$ $(t \in \mathcal{R})$ and consider the analytic continuation of the function $w(t)$ from the real axis into the complex plane

$$w(z) = z + iz^3 , \quad z \in \mathcal{C}. \tag{1.11.21}$$

Find critical points of the function (1.11.21). With the help of the program CONFORM create the mapping of the strip $-a < \mathrm{Im}\, z < a$ with $a = 1/\sqrt{6} \simeq 0.408$. Make sure the function (1.11.21) is univalent in this strip.

2. *Parabola of the fourth degree.* Represent the curve $v = u^4$ in the parametric form $w(t) = t + it^4$ $(t \in \mathcal{R})$ and consider the analytic continuation of the function $w(t)$ into the complex plane

$$w(z) = z + iz^4 , \quad z \in \mathcal{C}. \tag{1.11.22}$$

Find critical points of the function (1.11.22). Create the mapping of the strip $-a < \mathrm{Im}\, z < 0$ with $a = 2^{-2/3} \simeq 0.63$. Make sure the function (1.11.22) is univalent in this strip.

3. *Cosine curve.* Represent the curve $v = \cos u$ in the parametric form $w(t) = t + i \cos t$ $(t \in \mathcal{R})$ and consider the analytic continuation of the function $w(t)$ from the real axis into the complex plane

$$w(z) = z + i \cos z , \quad z \in \mathcal{C}. \tag{1.11.23}$$

Find critical points of the function (1.11.23). Create the mapping of the strip $-a < \mathrm{Im}\, z < a$ (with $a = \ln(1 + \sqrt{2}) \simeq 0.88$) and make sure the function (1.11.23) is univalent in this strip.

4. *Tangensoid.* Represent the curve $v = \arctan u$ in the parametric form $w(t) = t + i \arctan t$ $(t \in \mathcal{R})$ and consider the analytic continuation of the function $w(t)$ into the complex plane

$$w(z) = z + i \arctan z , \quad z \in \mathcal{C}. \tag{1.11.24}$$

The function (1.11.24) is analytic in the strip $-1 < \mathrm{Im}\, z < 1$. Create the mapping of the strip $-1 < \mathrm{Im}\, z < 1$ by this function and make sure it is univalent in this strip.

5. *The Gauss curve $v = e^{-u^2}$.* Create mappings of the Cartesian net in the strips $-\alpha < \mathrm{Im}\, z < 0$ and $0 < \mathrm{Im}\, z < \beta$ carried out by the function

$$w(z) = z + i e^{-z^2} , \quad z \in \mathcal{C}. \tag{1.11.25}$$

Make sure that for small enough α and β these mappings are univalent. Adjust as great as possible values α and β for which the function (1.11.25) preserves its univalence in these strips.

1.12 The Mapping of Polygonal Domains. Schwarz-Christoffel Integral

A singly connected domain of the extended complex plane bounded by a broken line is called the *polygonal domain*. All the polygonal domains are naturally separated into four types:

1. Finite polygons, i.e., domains lying inside closed finite broken lines. For such domains the point at infinity is the exterior one.

2. Infinite polygonal domains, that have one "vertex" at the point at infinity. For such domains the point $w = \infty$ is a one-fold boundary point. Such domains are called the *polygonal angular domains*.

3. Infinite polygonal domains, that have several vertices at the point at infinity. Such domains are called the *polygonal strips*. The point $w = \infty$ is the multi-fold boundary point for them.

4. Exteriors of finite closed broken lines, for that the point at infinity is the interior one. Such domains are called the *exteriors of polygons*.

Let us denote vertices of the n-gonal domain as $A_1, A_2, \ldots, A_n$, numbering them in the order of positive sense of the boundary traversal of the polygonal domain. The values of angles of the polygonal domain at these points are designated as $\pi\alpha_1, \pi\alpha_2, \ldots, \pi\alpha_n$ correspondingly. It can be considered without loss of generality that all $\alpha_k \neq 1$ (if some angle of the polygon is equal to π, it can be considered not to be as a vertex). The angles of the polygonal domain at the finite points A_k are considered to be positive, they satisfy the condition $0 < \alpha_k \leq 2$. The angles at the point at infinity are considered as negative, they satisfy the condition $-2 \leq \alpha_k \leq 0$.

The sum of interior angles of the finite n-gon (type 1) is equal to $\pi(n-2)$, i.e.,

$$\sum_{k=1}^{n} \alpha_k = n - 2 \,.$$

This formula is also valid for the infinite polygons (type 2 and 3). If the polygonal domain is the exterior of the finite n-gon (type 4), then

$$\sum_{k=1}^{n} \alpha_k = n + 2 \,.$$

Any polygonal domain of the w-plane can be conformally mapped onto the half-plane Im $z > 0$ in accordance with the Riemann theorem. Let the function $z = F(w)$ realize such a mapping and let the points of the real axis $a_1, a_2, \ldots, a_n$ be images of vertices $A_1, A_2, \ldots, A_n$ in the mapping. Let us denote as $w = f(z)$ the analytic function which realizes the inverse mapping of the upper half-plane onto the considered n-gonal domain ($f = F^{-1}$), in which the points $a_1, a_2, \ldots, a_n$ are mapped into the vertices of the polygonal domain $A_1, A_2, \ldots, A_n$. For finite or infinite polygons (types 1, 2 and 3) the mapping function has the form of an integral called Schwarz-Christoffel integral. In the case when all the points $a_1, a_2, \ldots, a_n$ (preimages of the polygon vertices) are finite, the Schwarz-Christoffel formula takes the form

$$f(z) = C_1 \int_{z_0}^{z} h(\xi)\, d\xi + C_2, \qquad (1.12.1)$$

where C_1, C_2, z_0 are complex constants,

$$h(\xi) = (\xi - a_1)^{\alpha_1 - 1}(\xi - a_2)^{\alpha_2 - 1} \cdots (\xi - a_n)^{\alpha_n - 1} \qquad (1.12.2)$$

is the function, which can be briefly written as

$$h(\xi) = \prod_{k=1}^{n}(\xi - a_k)^{\alpha_k - 1}. \qquad (1.12.3)$$

The function $h(\xi)$ is not single-valued in the general case, its branch points are the points $a_1, a_2, \ldots, a_n$ of the real axis. The principal branch of the function (1.12.3) is considered here, for which each multiplier $(\xi - a_k)^{\alpha_k - 1}$ is determined in the upper half-plane from the condition $0 < \arg(\xi - a_k) < \pi$.

In the case when one of the points a_k is the point at infinity (for example, $a_n = \infty$), the mapping of the upper half-plane onto the finite or infinite polygon is also defined by the integral (1.12.1), but the $h(\xi)$ expression includes $n - 1$ multipliers:

$$h(\xi) = \prod_{k=1}^{n-1}(\xi - a_k)^{\alpha_k - 1}, \qquad (1.12.4)$$

(i.e., the multiplier which corresponds to the value $a_n = \infty$ is absent). The constant z_0 in the integral (1.12.1) is insignificant and in most of the cases it can be taken equal to 0 (if this does not lead to the divergence of the integral).

The given polygonal domain of the w-plane can be conformally mapped onto the unit circle $|z| < 1$. Let the vertices of the polygon be mapped into the points of the unit circle $a_1, a_2, \ldots, a_n$. For finite or infinite polygons (types 1, 2 and 3) the inverse mapping of the circle onto the considered polygon is carried out by the function (1.12.1), (1.12.3) (but here $a_1, a_2, \ldots, a_n$ designate the points of the unit circumference).

For the mapping of the exterior of the polygon formulae (1.12.1–1.12.3) are not applicable. For this case the mapping of the exterior of the unit circle onto the exterior of the polygon is carried as follows. If angles of the infinite polygonal domain (i.e., the exterior angles of the polygon) equal to $\pi\alpha_1, \pi\alpha_2, \ldots, \pi\alpha_n$ and the points $a_1, a_2, \ldots, a_n$ of the unit circumference are preimages of the polygon vertices, the mapping function is

$$f(z) = C_1 \int_{z_0}^{z} g(\xi)\, d\xi + C_2\,, \qquad (1.12.5)$$

where

$$g(\xi) = \frac{1}{\xi^2} \prod_{k=1}^{n} (\xi - a_k)^{\alpha_k - 1}\,.$$

The function (1.12.5) maps the (interior) point $z = \infty$ into the point $w = \infty$.

When solving the problem on the conformal mapping of a half-plane or a circle onto the given polygonal domain, the parameters $a_1, a_2, \ldots, a_n$ and C_1, C_2 involved in formulae (1.12.1), (1.12.3) are unknown. The problem on the determination of these parameters is rather complicated and can not be solved analytically in the general form. The number of parameters to be determined can be reduced taking into account that according to the theorem of uniqueness of the conformal mapping (Subsection 1.6.3) preimages of three boundary points of a domain can be given arbitrarily. Thus in the problem on the mapping of triangles (see Subsection 1.12.3) the parameters of the function (1.12.1) are determined analytically and for polygons with greater number of sides the parameters can be determined analytically only in some particular cases. Only one of the parameters a_1, $a_2, \ldots, a_n$ can be arbitrarily given in the expression (1.12.5).

1.12.1 Star-shaped polygonal domains

Cuts consisting of segments and (or) rays of straight lines passing through the same point of a plane are called *star-shaped cuts*. The origin will be considered as this point. A singly connected domain bounded by a star-shaped cut is called a *star-shaped polygonal domain* or briefly — a star-shaped polygon. An analytic function which maps a half-plane or a circle onto a star-shaped polygon is an elementary one, it can be found without integration by the Schwarz-Christoffel formula.[106]

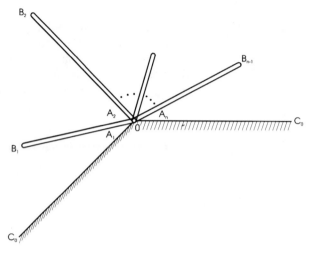

FIGURE 1.66

Angular star-shaped domains

Let the boundary of an angular star-shaped domain consist of two rays and $n - 1$ segments starting from the origin (Fig. 1.66). The origin is an n-folded boundary point. Designate the vertices of the polygonal domain at the origin (in the order of the positive sense of the boundary traversal) as A_1, A_2,...,A_n, and corresponding angles — as $\pi\alpha_1$, $\pi\alpha_2$,...,$\pi\alpha_n$. The point at infinity is a single-folded boundary point, an angle at the point at infinity C_0 is equal to

$$-\pi(\alpha_1 + \alpha_2 + \cdots + \alpha_n).$$

Designate the ends of segments as B_1, B_2,...,B_{n-1}, and the lengths of the segments $A_k B_k$ as l_k: $l_k = |A_k B_k|$. The side $A_n C_0$ is directed along the positive real axis.

The considered star-shaped domain can be mapped conformally onto the half-plane $\mathrm{Im}\, z > 0$ so that the point C_0 would be mapped into the point $z = \infty$. Let the points A_1, A_2,...,A_n be mapped into the points of the real axis a_1, a_2,...,a_n respectively. The inverse mapping of the half-plane $\mathrm{Im}\, z > 0$ onto the angular star-shaped domain is carried out by the function*

$$f(z) = C \prod_{k=1}^{n} (z - a_k)^{\alpha_k} . \tag{1.12.6}$$

The points b_k (preimages of the ends of the rays of the star B_k) are zeros of the derivative of the function $f(z)$, they are determined as roots of the equation

$$\frac{f'(z)}{f(z)} = \sum_{k=1}^{n} \frac{\alpha_k}{z - a_k} = 0 . \tag{1.12.7}$$

The equation (1.12.7) has exactly $n - 1$ real roots and

$$a_k < b_k < a_{k+1} \quad (k = 1, 2, \ldots, n - 1) .$$

When searching the mapping of the given domain the parameters a_1, $a_2, \ldots, a_n$ and C are unknown. To determine them $n - 1$ conditions are available

$$|f(b_k)| = l_k \quad (k = 1, 2, \ldots, n - 1) . \tag{1.12.8}$$

Two of the parameters a_1, $a_2, \ldots, a_n$ can be arbitrarily specified; the rest of them and the coefficient C are determined from the conditions (1.12.8).

Example 1.18

For the star-shaped angular domain with angles at the origin $\pi\theta$ and $2\pi - \pi\theta$ $(n = 2)$ we can assume

$$a_1 = 0, \quad a_2 = 1, \quad \alpha_1 = 2 - \theta, \quad \alpha_2 = \theta.$$

---

*As the formula (1.12.6) is not well known, we give the scheme of its proof.

The mapping function $f(z)$ in accordance with the symmetry principle can be analytically continued through some segment (a_k, a_{k+1}) into the lower half-plane and then through another segment (a_m, a_{m+1}) again into the upper half-plane. Analytically continued function maps the considered star-shaped polygon onto the same polygon, which is turned about the origin on some angle. Therefore, the analytically continued function $f(z)$ has the form $Cf(z)$. The function $f'(z)/f(z)$ is univalent. Its singular points are the points $a_1, a_2, \ldots, a_n$ where the function $f'(z)/f(z)$ has poles of the first order with residues $\alpha_1, \alpha_2, \ldots, \alpha_n$. Form the difference of the function $f'(z)/f(z)$ and the principal parts of the Laurent series at its poles:

$$\frac{f'(z)}{f(z)} - \sum_{k=1}^{n} \frac{\alpha_k}{z - a_k} = g(z) .$$

The function $g(z)$ is analytic in all the complex plane, it tends to 0 as $z \to \infty$. Due to the Liouville's theorem the function $g(z)$ identically equals 0, hence

$$\frac{f'(z)}{f(z)} = \sum_{k=1}^{n} \frac{\alpha_k}{z - a_k} .$$

The integration of this equality leads to the formula (1.12.6).

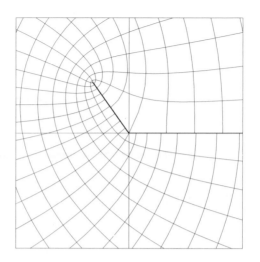

FIGURE 1.67

The formula (1.12.6) takes the form

$$w = Cz^{2-\theta}(z-1)^{\theta}. \tag{1.12.9}$$

The mapping of the Cartesian net in the half-plane Im $z > 0$ carried out by the function (1.12.9) with $\theta = 0.7$ is presented in the Fig. 1.67.

In the Fig. 1.68 the mapping of the half-plane Im $z > 0$ onto the symmetrical star-shaped domain $(n = 3)$ carried out by the function

$$w = Cz^{2-2\theta}(z+1)^{\theta}(z-1)^{\theta}$$

is presented also when $\theta = 0.7$.

Star-shaped strips with the origin being a boundary point

Consider a star-shaped polygonal strip with the origin being an n-folded boundary point and the point at infinity being an $m + 1$-folded one (Fig. 1.69).

Designate as A_1, A_2,...,A_n the boundary points of the star-shaped domain at the origin, the corresponding angles — as $\pi\alpha_1$, $\pi\alpha_2$,...,$\pi\alpha_n$. Further designate the boundary points at the point at infinity as C_0, C_1,...,C_m (also in the order of positive traversal of the domain boundary) and corresponding angles as $-\pi\gamma_0$, $-\pi\gamma_1$,...,$-\pi\gamma_m$. The angles of the star-shaped

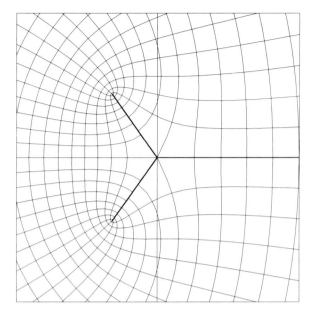

FIGURE 1.68

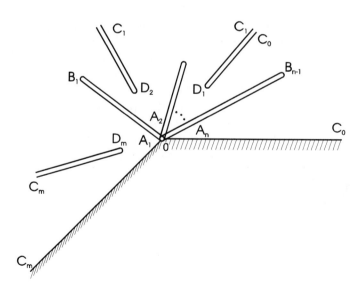

FIGURE 1.69

strip satisfy the relation

$$\sum_{k=1}^{n} \alpha_k = \sum_{k=0}^{m} \gamma_k \, .$$

The side $A_n C_0$ is directed along the positive real half-axis.

The considered strip can be conformally mapped onto the half-plane Im $z > 0$ so that the point C_0 would be mapped into the point at infinity. Let the points C_1, $C_2, \ldots, C_m$ be mapped into the points c_1, $c_2, \ldots, c_m$ of the real axis and the points A_1, $A_2, \ldots, A_n$ — into the points a_1, $a_2, \ldots, a_n$ respectively. It is obvious, that the points c_k and a_k will be disposed in the order:

$$c_1 < c_2 < \ldots < c_m < a_1 < a_2 < \ldots < a_n \, .$$

The function realizing the inverse mapping of the half-plane Im $z > 0$ onto the considered star-shaped strip has the form

$$w = C \, \frac{\prod_{k=1}^{n}(z - a_k)^{\alpha_k}}{\prod_{k=1}^{m}(z - c_k)^{\gamma_k}} \, , \qquad (1.12.10)$$

where $C > 0$. When solving the problem of the mapping of a half-plane onto the given polygon the parameters a_k, c_k are unknown. Two of them can be arbitrarily chosen and the rest of them and the coefficient C (totally $n + m - 1$ parameters) are found from the system of $n + m - 1$ equations, which determines the position of the points $B_1, \ldots, B_{n-1}$; $D_1, \ldots, D_m$ (Fig. 1.69). Practically (for example, when using the program CONFORM) instead of the solution of the complicated system of equations the method of parameters fitting is used to obtain the necessary shape of the domain.

The case of the strip with two branches ($m = 1$) is particularly significant. Such strips can be mapped onto the adequate canonical domain — the strip $0 < $ Im $z < \pi$ — in such a manner that the boundary points C_1 and C_0 of the star-shaped strip would be mapped into the points $\xi = -\infty$ and $\xi = \infty$ respectively. This can be achieved by the transformation $\xi = \ln(z - c_1)$. The inverse mapping of the strip $0 < $ Im $\xi < \pi$ onto the star-shaped strip is realized by the composition of functions

$$z = c_1 + e^{\xi} \, ,$$

$$w = \frac{C}{(z - c_1)^{\gamma_1}} \prod_{k=1}^{n}(z - a_k)^{\alpha_k} \, .$$

Example 1.19

The mapping of the Cartesian net in the strip $0 < $ Im $z < \pi$ realized by

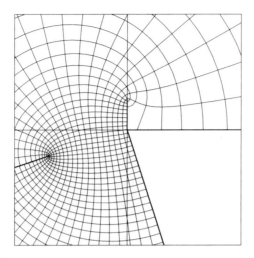

FIGURE 1.70

the function

$$w = \frac{(z_1 - 1)^{\alpha_1}(z_1 - p)^{\alpha_2}}{z_1^{\gamma}}$$

is presented in Fig. 1.70 ($\alpha_1 = 1.1$, $\alpha_2 = 0.5$, $\gamma = 0.5$, $p = 5$, $z_1 = e^z$).

Star-shaped polygonal domains with the origin being an internal point

Let the boundary of a star-shaped domain consist of m cuts along straight line rays with continuations passing through the origin. The point $w = 0$ is the internal point for such a domain and infinity is an m-folded boundary point. Designate the boundary points at infinity as C_1, $C_2, \ldots, C_m$ and the angles at these points as $-\pi\gamma_1$, $-\pi\gamma_2, \ldots, -\pi\gamma_m$ respectively (all $\gamma_k > 0$, $\sum_{k=1}^{m} \gamma_k = 2$) (Fig. 1.71).

Consider a conformal mapping of the given star-shaped strip with m branches onto the unit circle $|z| < 1$, the point $w = 0$ is mapped into the center of the circle. Let the points $C - 1$, $C_2, \ldots, C_m$ be mapped into the points of the unit circumference $c_1 = e^{i\varphi_1}$, $c_2 = e^{i\varphi_2}, \ldots, c_m = e^{i\varphi_m}$. The function which realizes the inverse mapping of the unit circle onto the star-shaped domain has the form

$$w = \overline{C}z \prod_{k=1}^{m} (z - c_k)^{-\gamma_k},$$

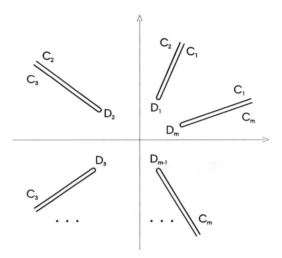

FIGURE 1.71

where $\overline{C}$ is some complex constant. This function is multi-valued and breaks appear at points of some segments lying inside the circle in the process of extraction of the principal branches of its factors. To avoid these breaks it is necessary to present the mapping function in the form

$$w = Cz \prod_{k=1}^{m} \left(1 - \frac{z}{c_k}\right)^{-\gamma_k}. \tag{1.12.11}$$

The parameters $c_1, c_2, \ldots, c_m$ are unknown. Only one of them can be arbitrarily preassigned (for example, $c_1 = 1$).

A particular case of the formula (1.12.11) is the formula (1.10.13). It can be derived by substituting into (1.12.11)

$$\gamma_k = 2\pi/m, \quad c_k = e^{i\pi(2k+1)/m}, \quad C = 2^{2/m},$$

where $k = 1, 2, \ldots, m$.

Exteriors of finite star-shaped cuts

Let an infinite star-shaped domain be the exterior of a star consisting of n segments, starting from the point $w = 0$ (Fig. 1.72). Designate the vertices in the origin (in order of the positive traversal of the boundary of the infinite domain) as $A_1, A_2, \ldots, A_n$ and the angles at these vertices as $\pi\alpha_1, \pi\alpha_2, \ldots, \pi\alpha_n$ correspondingly ($\sum_{k=1}^{n} \alpha_k = 2$).

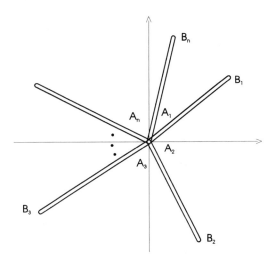

FIGURE 1.72

The exterior of the star-shaped cut can be conformally mapped onto the adequate canonical domain (the exterior of the unit circle $|z| > 1$), the point $w = \infty$ being mapped into the point $z = \infty$. Let the points A_1, $A_2, \ldots, A_n$ be mapped into the points $a_1, a_2, \ldots, a_n$ of the unit circumference, correspondingly.

The function realizing the inverse mapping of the exterior of the unit circle onto the exterior of the star-shaped cut has the form

$$w = \frac{C}{z} \prod_{k=1}^{n} (z - a_k)^{\alpha_k}. \tag{1.12.12}$$

To extract continuous branch of this multi-valued function outside the circle $|z| > 1$ the expression (1.12.12) should be represented as

$$w = Cz \prod_{k=1}^{n} \left(1 - \frac{a_k}{z}\right)^{\alpha_k}. \tag{1.12.13}$$

The formula (1.12.13) can be obtained from (1.12.11) by inversions on z and w.

The parameters $a_1, a_2, \ldots, a_n$ are unknown. One of them can be arbitrarily preassigned, for example, $a_1 = 1$. The rest of the parameters $a_k = e^{i\varphi_k}$ are necessary to define (or to fit) in order to obtain the given star-shaped cut.

A particular case of the formula (1.12.13) is the formula (1.10.12). It can be derived by substituting

$$\alpha_k = 2\pi/n, \quad a_k = e^{i\pi(2k+1)/n}, \quad C = 2^{-2/n}$$

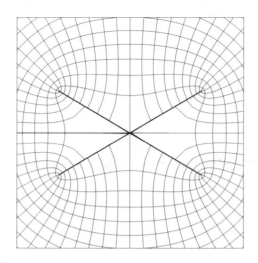

FIGURE 1.73

into the formula (1.12.13).

Example 1.20

Consider the star-shaped cut, consisting of four segments, coming from the origin and being disposed symmetrically with respect to coordinate axis. Let the angles between the segments be equal to πp, $\pi(1-p)$, πp, $\pi(1-p)$. It is natural to take symmetrical points 1, $-i$, -1, i on the unit circumference as the points a_k. The formula (1.12.12) yields

$$w = \frac{C}{z}(z-1)^p(z+i)^{1-p}(z+1)^p(z-i)^{1-p} = C\left(z+\frac{1}{z}\right)\left(\frac{z^2-1}{z^2+1}\right)^p.$$
$$(1.12.14)$$

The mapping of the exterior of the circle $|z| > 1$ realized by the function (1.12.14) with $p = 1/3$ is presented in Fig. 1.73.

_____

Exercises

1. Create the mapping of the Cartesian net in the upper half-plane Im $z > 0$ realized by the function

$$w = i(a^2 - z^2)^p(1 - z^2)^{1/2-p}$$

   with $a = 2$, $p = 0.3$. Examine experimentally the changes of the ratio of lengths of cuts under the changes of the parameter a ($a > 1$).

2. For the star-shaped domain bounded by four rays symmetrical with respect

to the coordinate axis the function (1.12.11) takes the form

$$w = \frac{Cz}{(1 - z^2)^p (1 + z^2)^{1-p}}$$ (1.12.15)

after the substitution

$$c_1 = 1, \qquad c_2 = i, \qquad c_3 = -1, \qquad c_4 = -i;$$
$$\gamma_1 = p, \qquad \gamma_2 = 1 - p, \qquad \gamma_3 = p, \qquad \gamma_4 = 1 - p.$$

Create the mapping of the unit circle realized by the function (1.12.15) with $p = 0.6$.

3. Create the mapping of the Cartesian net in the strip $-\pi/2 < \mathrm{Re}\, z < \pi/2$ realized by the function

$$w = \sin z (\cos z)^{-p}$$ (1.12.16)

with $p = 0.6$.

   Note. The function (1.12.16) is obtained from (1.12.15) by substituting z for $\tan z/2$ and to assign $C = 2$.

4. Create the mapping of the exterior of the circle $|z| > 1$ realized by the function

$$w = z \left(1 - \frac{1}{z}\right)^{2p} \left[\left(1 - \frac{e^{i\pi\varphi}}{z}\right)\left(1 - \frac{e^{-i\pi\varphi}}{z}\right)\right]^{1-p}$$

with $p = 0.333$, $\varphi = 0.667$. Examine experimentally the change of lengths of sides of the cut with the change of the value φ (to take values $\varphi = 0.5$ and $\varphi = 0.8$).

1.12.2   Polygons bounded by rays of parallel straight lines

In the cases when the boundary of a singly connected polygonal domain consists of parallel rays and, probably, of one or two straight lines parallel to them, the Schwarz-Christoffel integral is expressed as elementary functions. At that a part of the parameters in the Schwarz-Christoffel formula is expressed through geometrical characteristics of the domain.[164] All the considered domains are strips with several branches.

Infinite plane with cuts along parallel rays having the same direction

Let the boundary of the domain G on the plane w consist of n cuts along the rays that are parallel to the real axis and directed along the negative real half-axis (Fig. 1.74).

   Denote as $h_1, h_2, \dots, h_{n-1}$ distances between neighboring cuts, as B_1, $B_2, \dots, B_n$ — the ends of rays (in the order of the positive traversal of the boundary of the domain), as $A_0, A_1, \dots, A_{n-1}$ — the points of the boundary of the domain lying at infinity. Suppose the lower cut $A_{n-1} B_n A_n$ to lie on the real axis.

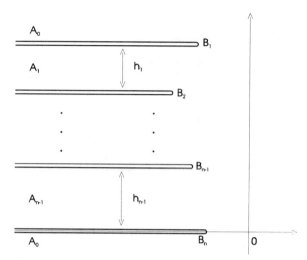

FIGURE 1.74

According to the Riemann's theorem the considered domain G can be mapped onto the half-plane Im $z > 0$, the point A_0 being mapped into the point $z = \infty$. Let the points $A_1, A_2, \ldots, A_{n-1}$ be mapped into the points of the real axis $a_1, a_2, \ldots, a_{n-1}$, the points $B_1, B_2, \ldots, B_n$ — into the points $b_1, b_2, \ldots, b_n$ correspondingly. The inverse mapping of the half-plane Im $z > 0$ onto the domain G is realized by the function (1.12.1)

$$w = \overline{C_1} \int_{z_0}^{z} h(\xi)\, d\xi + \overline{C_2},$$

where

$$h(\xi) = \frac{\displaystyle\prod_{k=1}^{n} (\xi - b_k)}{\displaystyle\prod_{k=1}^{n-1} (\xi - a_k)}. \qquad (1.12.17)$$

The fraction (1.12.17) is improper. Extracting its integral part and decomposing it into partial fractions we have

$$h(\xi) = \xi + \alpha_0 + \sum_{k=1}^{n-1} \frac{\alpha_k}{\xi - a_k},$$

where $\alpha_0, \alpha_1, \ldots, \alpha_{n-1}$ are real coefficients.

The Schwarz-Christoffel integral takes the form

$$w = \overline{C_1} \left[\frac{z^2}{2} + \alpha_0 z + \sum_{k=1}^{n-1} \alpha_k \ln(z - a_k) \right] + \overline{C_2}. \qquad (1.12.18)$$

For the function (1.12.18) to be real-valued it is necessary for the coeffi-cients $\overline{C_1}$ and $\overline{C_2}$ to be real.

During the traversal of the point $z = a_k$ along a small half-circumference in the upper half-plane the imaginary part of the function $\ln(z - a_k)$ gets the increment $-\pi$. The increment of the function w should be equal to $-ih_k$, thus $\overline{C_1}\alpha_k = h_k/\pi$. Let us denote $\overline{C_1}\alpha_0 = B$, $\overline{C_1}/2 = -A$, $\overline{C_2} = C$. Finally the formula (1.12.18) yields

$$w = -Az^2 + Bz + C + \frac{1}{\pi}\sum_{k=1}^{n-1} h_k \ln(z - a_k). \qquad (1.12.19)$$

All the parameters in the formula (1.12.19) are real-valued and $A > 0$, $h_k > 0$. Two of the parameters $a_1, a_2,\ldots,a_{n-1}$ can be arbitrarily preas-signed. The rest $n - 3$ parameters and coefficients A, B, C are defined by n conditions

$$w(b_k) = B_k, \quad k = 1, 2, \ldots, n,$$

where b_k are zeros of the derivative of (1.12.19):

$$w' = -2Az + B + \frac{1}{\pi}\sum_{k=1}^{n-1}\frac{h_k}{z - a_k} = 0.$$

Practically the parameters a_k (when $n > 3$) are defined by fitting.

Example 1.21
When $n = 2$, $A = 1$, $B = 0$, $h = 2\pi$, $C = 1 - i\pi$, $a_1 = 0$ the func-tion (1.12.19) has the form

$$w = -z^2 + 2\ln z - i\pi + 1.$$

The mapping of the Cartesian net in the half-plane $\operatorname{Im} z > 0$ realized by the function is presented in Fig. 1.75.

When $n = 2$ the domain depicted in Fig. 1.74 is a curvilinear strip. The conformal mapping of the adequate canonical domain being the straight line strip $0 < \operatorname{Im} \xi < \pi$ onto the domain under consideration is realized by the function

$$w = -Az^2 + Bz + C + \frac{h}{\pi}\ln z, \qquad (1.12.20)$$

where $z = e^{\xi}$.

If $B = 0$ the function (1.12.20) differs from the Kirchhoff's func-tion (1.11.16) only by the linear transformation of an independent variable.

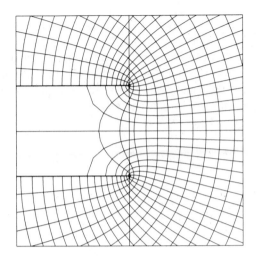

FIGURE 1.75

Example 1.22

The mapping of the Cartesian net in the strip $0 < \operatorname{Im} \xi < \pi$ realized by
the function (1.12.20) with $A = 1$, $B = 1$, $C = 1 - i\pi$, $h = 2\pi$ is presented
in Catalog 2 (Domain 20).

Plane with cuts along parallel oppositely directed rays

Let the domain G be an infinite plane w with $n + m$ cuts along the rays
parallel to the real axis, n of them being directed toward negative values
of x and m rays — toward the positive values of x (Fig. 1.76). Let the cut
$C_{m-1} D_m A_0$ lie on the real axis.

Designate the ends of the rays as $B_1, B_2, \ldots, B_n$; $D_1, D_2, \ldots, D_m$, boundary points at infinity (in the order of the positive traversal of the boundary)
— as $A_0, A_1, \ldots, A_{n-1}$; $C_0, C_1, \ldots, C_{m-1}$.

The domain G can be conformally mapped onto the half-plane $\operatorname{Im} z >$
0, the point A_0 being mapped into the point $z = \infty$. Let the points
A_k, B_k, C_k, D_k be mapped into the points of the real axis a_k, b_k, c_k,
d_k correspondingly. The Schwarz-Christoffel integral for the considered
domain takes the form

$$w(z) = \overline{C_1} \int_{z_0}^{z} h(\xi)\, d\xi + \overline{C_2},$$

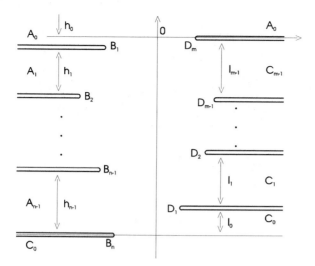

FIGURE 1.76

where

$$h(\xi) = \frac{\prod\limits_{k=1}^{n}(\xi - b_k)\prod\limits_{k=1}^{m}(\xi - d_k)}{\prod\limits_{k=1}^{n-1}(\xi - a_k)\prod\limits_{k=1}^{m-1}(\xi - c_k)}\,\frac{1}{(\xi - c_0)^2}\,. \qquad (1.12.21)$$

The fraction $h(\xi)$ is improper. Extracting its integral part and decomposing it into partial fractions we have

$$h(\xi) = 1 + \frac{\alpha_0}{(\xi - c_0)^2} + \sum_{k=1}^{n-1}\frac{\alpha_k}{\xi - a_k} + \sum_{k=0}^{m-1}\frac{\beta_k}{\xi - c_k}\,,$$

where α_0, α_k, β_0, β_k are real coefficients. The function mapping the half-plane Im $z > 0$ onto the considered domain is

$$w = \overline{C_1}\left[z - \frac{\alpha_0}{z - c_0} + \sum_{k=1}^{n-1}\alpha_k\ln(z - a_k) + \sum_{k=0}^{m-1}\beta_k\ln(z - c_k)\right] + \overline{C_2}\,.$$

Since the function $\ln(z - a_k)$ gets the increment $-i\pi$ during the traversal of the point $z = a_k$ along a small upper half-circumference, the coefficients are expressed as

$$\alpha_k\overline{C_1} = \frac{h_k}{\pi}\,, \quad k = 1, 2, \ldots, n - 1\,.$$

In a similar way β_k coefficients are defined: $\beta_k\overline{C_1} = -l_k/\pi$ ($k = 1$, $2, \ldots, m - 1$). Finally, designating $\overline{C_1}\alpha_0 = -B$, $\overline{C_1} = A$, $\overline{C_2} = C$ we can

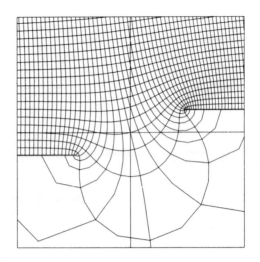

FIGURE 1.77

represent the mapping function $w(z)$ as

$$w(z) = Az + \frac{B}{z - c_0} + C + \frac{1}{\pi} \sum_{k=1}^{n-1} h_k \ln(z - a_k) - \frac{1}{\pi} \sum_{k=0}^{m-1} l_k \ln(z - c_k). \quad (1.12.22)$$

The parameters A, B, h_k $(k = 1, 2, \ldots, n-1)$ and l_k $(k = 1, 2, \ldots, m-1)$ should be positive, coefficients l_0 and C can be arbitrary real numbers. The real parameters a_k and c_k form the growing sequence

$$a_1 < a_2 < \ldots < a_{n-1} < c_0 < c_1 < \ldots < c_{m-1}.$$

Two of these parameters can be arbitrarily preassigned.

Example 1.23
If $n = 1$, $m = 1$, $c_0 = 0$ the function (1.12.22) yields

$$w = Az + \frac{B}{z} + C - \frac{l_0}{\pi} \ln z. \quad (1.12.23)$$

The mapping of the Cartesian net in the half-plane $\operatorname{Im} z > 0$ realized by the function (1.12.23) with $A = B = 1$, $l_0 = \pi/2$, $C = i\pi/4$ is presented in Fig. 1.77.

Similar to the Examples 1.21 and 1.22 the domain depicted in Fig. 1.77 is a curvilinear strip. The mapping of an adequate canonical domain — the domain $0 < \operatorname{Im} \xi < \pi$ onto the considered domain is carried out by the function (1.12.23), where $z = e^\xi$ should be assumed.

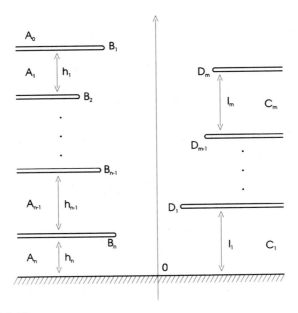

FIGURE 1.78

Example 1.24

The mapping of the Cartesian net in the strip $0 < \text{Im } \xi < \pi$ realized by the function (1.12.23) with the substitution $z = e^\xi$ is presented in the Catalog 4, Domain 19. Another solution of this problem is given in Subsection 1.11.7.

Half-plane with cuts along parallel rays

Let the domain G on the plane w be the half-plane $\text{Im } w > 0$ with $n + m$ cuts made along the rays parallel to the real axis, n rays start from the points B_k ($k = 1, 2,\ldots,n$) to the left and m rays start from the points D_k ($k = 1, 2,\ldots,m$) to the right (Fig. 1.78).

Designate boundary points lying at infinity as A_0, $A_1,\ldots,A_n$; C_1, $C_2,\ldots,C_m$ (in the order of the positive traversal of the boundary). The considered domain can be conformally mapped onto the half-plane $\text{Im } z > 0$, the point A_0 being mapped into the point $z = \infty$. Let the points A_k, B_k, C_k, D_k be mapped into the points of the real axis a_k, b_k, c_k, d_k correspondingly. The inverse mapping of the half-plane $\text{Im } z > 0$ onto the considered polygonal domain is carried out by the function (1.12.1) with

$$
h(\xi) = \frac{\prod\limits_{k=1}^{n} (\xi - b_k) \prod\limits_{k=1}^{m} (\xi - d_k)}{\prod\limits_{k=1}^{n} (\xi - a_k) \prod\limits_{k=1}^{m} (\xi - c_k)} \, . \tag{1.12.24}
$$

Extracting the integral part from the fraction $h(\xi)$ and decomposing it into partial fractions we have

$$
h(\xi) = 1 + \sum_{k=1}^{n} \frac{\alpha_k}{\xi - a_k} + \sum_{k=0}^{m} \frac{\beta_k}{\xi - c_k} \, ,
$$

where α_k, β_k are real coefficients.

The Schwarz-Christoffel integral yields

$$
w = \overline{C_1} \left[z + \sum_{k=1}^{n} \alpha_k \ln(z - a_k) + \sum_{k=0}^{m} \beta_k \ln(z - c_k) \right] + \overline{C_2} \, .
$$

The functions $\ln(z - a_k)$ and $\ln(z - c_k)$ get the increment $-i\pi$ during the traversal of the points $z = a_k$ and $z = c_k$ along a small half-circumferences. Hence

$$
\alpha_k \overline{C_1} = \frac{h_k}{\pi} \, , \qquad \beta_k \overline{C_1} = -\frac{l_k}{\pi} \, .
$$

Designate $\overline{C_1} = A$, $\overline{C_2} = C$ and represent the mapping function as

$$
w = Az + C + \frac{1}{\pi} \sum_{k=1}^{n} h_k \ln(z - a_k) - \frac{1}{\pi} \sum_{k=1}^{m} l_k \ln(z - c_k) \, . \tag{1.12.25}
$$

The coefficients A, h_k, l_k are positive in the formula (1.12.25). The coefficient C satisfies the condition $\operatorname{Im} C = \sum l_k$. The parameters a_k and c_k satisfy the conditions

$$
a_1 < a_2 < \ldots < a_n < c_1 < c_2 < \ldots < c_m \, .
$$

Two of these parameters can be arbitrarily preassigned.

Example 1.25

In the simplest case ($n = 1$, $m = 0$) the domain G is the half-plane with the single cut. In this case the formula (1.12.25) yields

$$
w = Az + \frac{h}{\pi} \ln z + C \, . \tag{1.12.26}
$$

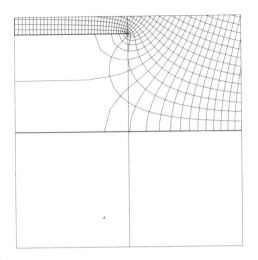

FIGURE 1.79

The change of the variable $\pi A z / h = e^{\xi}$ turns this function into the Kirchhoff's function (1.11.16)

$$w = \frac{h}{\pi}(\xi + e^{\xi}) + \overline{C}.\qquad(1.12.27)$$

The mapping of the Cartesian net in the half-plane Im $z > 0$ carried out by the function (1.12.26) with $A = 1$, $h = \pi$, $C = 1$ is presented in Fig. 1.79.

The mapping of the Cartesian net in the strip $0 < \text{Im}\,\xi < \pi$ realized by the function (1.12.27) with $h = \pi$, $\overline{C} = 1$ is presented in Fig. 1.80.

In the case $n + m > 2$ the domain depicted in Fig. 1.78 is a strip with three or more branches. In this case the considered domain can be mapped onto the straight line strip $0 < \text{Im}\,z < \pi$ by various means. See Section 2.13 for more details concerning the mapping of strips with several branches.

Example 1.26

If $n = 1$, $m = 1$ it is possible to assume $a_1 = -1$, $c_1 = 1$ in the formula (1.12.25). The function mapping the half-plane Im $z > 0$ onto the considered domain has the form

$$w = Az + C + \frac{h}{\pi}\ln(z+1) - \frac{l}{\pi}\ln(z-1).\qquad(1.12.28)$$

The mapping of the Cartesian net in the upper half-plane realized by the function (1.12.28) with $h = 1.4\pi$, $l = \pi$, $A = 1$, $C = 0$ is presented in Fig. 1.81.

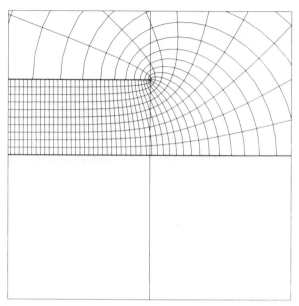

FIGURE 1.80

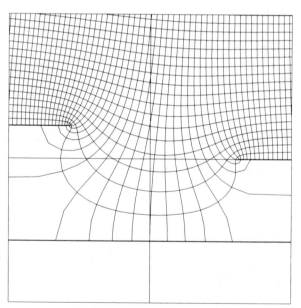

FIGURE 1.81

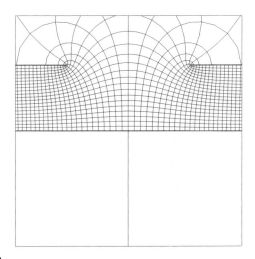

FIGURE 1.82

Another mapping of the considered domain can be obtained by the construction of the mapping of the strip $0 < \operatorname{Im}\xi < \pi$ onto the half-plane $\operatorname{Im} z > 0$, the points $\xi = \pm\infty$ being mapped into the points $z = \pm 1$. Such a mapping is carried out by the function $z = \tanh(\xi/2)$. The mapping of the Cartesian net in the strip $0 < \operatorname{Im}\xi < \pi$ realized by the function (1.12.28), where $z = \tanh(\xi/2)$, $A = 1$, $C = 0$, $h = 1.4\pi$, $l = \pi$ is presented in Catalog 5, Domain 3.

The function (1.12.28) has an especially simple form if $h = l = \pi$, $C = i\pi$:

$$w = Az + \ln\frac{1+z}{1-z} = Az + 2\operatorname{arth} z.$$

After substitution $z = \tanh(\xi/2)$ the function yields

$$w = A\tanh(\xi/2) + \xi. \qquad (1.12.29)$$

If $A > 0$ the function (1.12.29) maps the strip $0 < \operatorname{Im}\xi < \pi$ onto the half-plane $\operatorname{Im} w > 0$ with two cuts along symmetrical rays (Fig. 1.82).

If $A = -1$ the function (1.12.29) maps the strip $-\pi/2 < \operatorname{Im}\xi < \pi/2$ onto the domain between two tractrices (see Subsection 1.10.12).

Example 1.27
If $n = 2$, $m = 0$, $a_1 = -1$, $a_2 = 1$, $C = -ih_2$ the function (1.12.29) takes the form

$$w = Az + \frac{h_1}{\pi}\ln(1+z) + \frac{h_2}{\pi}\ln(1-z). \qquad (1.12.30)$$

Assuming $z = \tanh(\xi/2)$ we obtain the function, which maps the strip

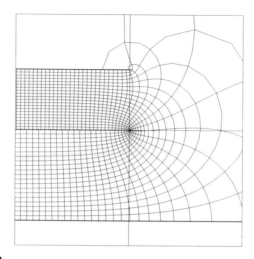

FIGURE 1.83

$0 < \operatorname{Im} \xi < \pi$ onto the strip with three branches. The mapping of the strip $0 < \operatorname{Im} \xi < \pi$ realized by the function (1.12.30) with $A = 0.9$, $h_1 = \pi$, $h_2 = 1.5\pi$ is presented in Fig. 1.83.

Strip with cuts along rays

Let the domain G represents the horizontal strip $0 < \operatorname{Im} w < H$ with $n + m - 2$ cuts along rays starting from the points B_k to the left ($k = 1$, $2,\ldots,n - 1$) and from the points D_k to the right ($k = 1$, $2,\ldots,m - 1$) (Fig. 1.84). Designate as A_1, $A_2,\ldots,A_n$; C_1, $C_2,\ldots,C_m$ boundary points lying at infinity and numbered in the order of the positive traversal of the boundary of the domain G. The domain G can be conformally mapped onto the half-plane $\operatorname{Im} z > 0$. Let the boundary points A_k, B_k, C_k, D_k be mapped into the points a_k, b_k, c_k, d_k of the real axis. Suppose that none of the points a_k, b_k, c_k, d_k is the point at infinity. The function realizing the inverse mapping of the half-plane $\operatorname{Im} z > 0$ onto the domain G is expressed through the Schwarz-Christoffel formula

$$w = \overline{C_1} \int_{z_0}^{z} h(\xi)\, d\xi + \overline{C_2}\,,$$

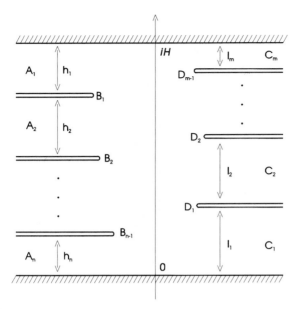

FIGURE 1.84

where

$$h(\xi) = \frac{\prod\limits_{k=1}^{n-1} (\xi - b_k) \prod\limits_{k=1}^{m-1} (\xi - d_k)}{\prod\limits_{k=1}^{n} (\xi - a_k) \prod\limits_{k=1}^{m} (\xi - c_k)}.$$

The proper fraction $h(\xi)$ is decomposed into the partial fractions

$$h(\xi) = \sum_{k=1}^{n} \frac{\alpha_k}{\xi - a_k} + \sum_{k=1}^{m} \frac{\beta_k}{\xi - c_k},$$

where α_k, β_k are real coefficients. After the integration we obtain

$$w = \overline{C_1} \left[\sum_{k=1}^{n} \alpha_k \ln(z - a_k) + \sum_{k=1}^{m} \beta_k \ln(z - c_k) \right] + \overline{C_2}.$$

The increments of the function $w(z)$ during the traversal of the points a_k and c_k should be equal to $-ih_k$ and il_k respectively. Since the increment of the function $\ln(z - a_k)$ during the traversal of the point a_k along the half-circumference lying in the upper half-plane is equal to $-i\pi$,

$$\overline{C_1}\alpha_k = \frac{h_k}{\pi}, \quad \overline{C_1}\beta_k = -\frac{l_k}{\pi}.$$

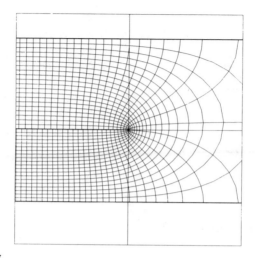

FIGURE 1.85

Thus the function mapping the half-plane Im $z > 0$ onto the domain G has the form

$$w = \frac{1}{\pi} \sum_{k=1}^{n} h_k \ln(z - a_k) - \frac{1}{\pi} \sum_{k=1}^{m} l_k \ln(z - c_k), \qquad (1.12.31)$$

where Im $\overline{C_2} = H = \sum l_k$. If one of the parameters a_k or c_k is equal to ∞, the corresponding term in the formula (1.12.31) is omitted.

Example 1.28

If $n = 2$, $m = 1$ the domain G is the strip $0 < $ Im $w < H$ with the cut along the ray. Assuming $a_1 = -1$, $a_2 = 1$, $c_1 = \infty$, $\overline{C_2} = -ih_2$; $z = \tanh(\xi/2)$ in the formula (1.12.31) we obtain the function, which maps the strip $0 < $ Im $\xi < \pi$ onto the considered curvilinear strip:

$$z = \tanh(\xi/2),$$

$$w = \frac{h_1}{\pi} \ln(1 + z) + \frac{h_2}{\pi} \ln(1 - z). \qquad (1.12.32)$$

The mapping of the Cartesian net in the strip $0 < $ Im $\xi < \pi$ realized by the function (1.12.32) with $h_1 = \pi$, $h_2 = 0.8\pi$ is presented in Fig. 1.85.

Exercises

1. The mapping of the strip $0 < \text{Im } z < \pi$ onto the plane with three cuts along codirected rays parallel to the real axis is carried out by the composition of functions

$$t = \tanh(z/2), \tag{1.12.33}$$

$$w = \ln(1+t) + \ln(1-t) - At^2. \tag{1.12.34}$$

The function (1.12.34) is a particular case of the function (1.12.19) if $n = 3$, $a_1 = -1$, $a_2 = 1$, $h_1 = h_2 = \pi$, $B = 0$, $C = -i\pi$. Create the mapping of the Cartesian net in the strip $0 < \text{Im } z < \pi$ realized by the function (1.12.33–1.12.34) with $A = 0.1, 0.2785, 0.5$.

2. The function

$$w(t) = t + 1/t + 2\ln(1+t) - \ln t \tag{1.12.35}$$

is a particular case of the function (1.12.22) if $A = B = 1$, $C = 0$, $n = 2$, $m = 1$, $h_1 = 2\pi$, $l_0 = \pi$, $a_1 = -1$, $c_0 = 0$. Create the mapping of the Cartesian net in the strip $0 < \text{Im } z < \pi$ realized by the function (1.12.35), where $t = e^z$.

3. The function

$$w = \ln(t+p) + q\ln(t+1) - q\ln(t-1) - \ln(p-t)$$

where $t = \tanh(z/2)$, $p > 1$, $q > 0$ maps the strip $0 < \text{Im } z < \pi$ onto the strip with two cuts along rays of the same straight line. Create the mapping carried out by the function with $p = 3$, $q = 1.2$.

4. The function

$$w = A/(t^2 - 1) + \ln t - i\pi/2,$$

where $t = e^z$, $A > 0$ maps the strip $0 < \text{Im } z < \pi$ onto the plane w with four cuts along rays, which are parallel to the real axis and disposed symmetrically with respect to the coordinate axis. Create the mapping of the Cartesian net carried out by the function.

1.12.3   The Mapping of triangular domains

In accordance with the general classification all triangular domains can be divided into four types:

1. finite triangles,

2. triangular angular domains with the point at infinity being the onefold boundary point,

3. triangular strips with the point $z = \infty$ being the two-fold boundary point,

4. exteriors of triangles.

Let A_1, A_2, A_3 be the vertices of the triangular domain numbered in the order of the positive traversal of the domain boundary, $\pi\alpha$, $\pi\beta$, $\pi\gamma$ be the values of angles at the points A_1, A_2, A_3 correspondingly. At least one of the points A_k is the finite one, suppose it to be A_1, therefore $\alpha > 0$.

For the mapping of finite and infinite triangles (of types 1, 2, 3) the formulae (1.12.1), (1.12.4) are true. Using the free choice of three parameters in the expression (1.12.1) let us take

$$a_1 = 0, \quad a_2 = 1, \quad a_3 = \infty.$$

The formula (1.12.1) is reduced to

$$w = C_1 \int_0^z \xi^{\alpha-1}(\xi - 1)^{\beta-1}\, d\xi + C_2. \qquad (1.12.36)$$

The expression $\xi^{\alpha-1}(\xi-1)^{\beta-1}\, d\xi$ is called the *binomial differential*. Russian mathematician P. L. Chebyshev has proved the following theorem concerning binomial differentials:[†]

THEOREM 1.18

If α and β are rational numbers, the integral (1.12.36) is expressed as elementary functions only when either α or β or $\alpha+\beta$ is an integer number.

The following Addition to the Chebyshev's theorem can be done. The integral (1.12.36) is also expressed as the elementary functions when one of the parameters α or β is a natural number and the other is an irrational one.

An angle of the polygonal domain being a multiple of π is called the *integer* one. Since angles of the polygonal domain satisfy the condition $-2 \leq \alpha_k \leq 2$, the integer angle can be equal to 2π (at the finite point) or 0, $-\pi$ or -2π (at the point at infinity). Angles of the finite or infinite triangle (of types 1–3) satisfy the condition $\alpha+\beta+\gamma = 1$, which yields $\gamma = 1-(\alpha+\beta)$. The Chebyshev's theorem and its Addition can be reformulated for the triangular domains as follows:

Schwarz-Christoffel integral (1.12.36) is expressed as the elementary functions only then, when one of the angles α, β, γ is the integer one. If the integer angle is equal to 2π, each of the other angles can be any of the real numbers, and if the integer angle is equal to 0, $-\pi$ or -2π, then the other angles should have a form p/q, where p and q are integer numbers. Let us call angles $\pi p/g$ (p, q are integer numbers) the *rational* ones.

The Chebyshev's theorem and its Addition permit us to consider all the possible cases, when the integral 1.12.36 is expressed as elementary functions.

[†]Chebyshev considered the binomial differential in more general form $x^m(ax^n - b)^p\, dx$, which can be reduced to the considered case by the substitution $z = ax^n/b$, $\alpha = (m + 1)/n$, $\beta = p + 1$.

First of all note the simplest case, when all the angles of the triangular domain are integer. It is possible only for the combination of angles $(2\pi, 0, -\pi)$. In this case the domain is the half-plane with the cut along the ray parallel to the edge of the half-plane. The conformal mapping of the domain was considered above in the Subsection 1.12.2 (Example 1.25).

Finite triangles

Angles of finite triangles satisfy the conditions $0 < \alpha < 1$, $0 < \beta < 1$, $0 < \gamma < 1$, $\alpha + \beta + \gamma = 1$. They do not satisfy the demands of the Chebyshev's theorem, therefore the half-plane Im $z > 0$ can not be conformally mapped onto the finite triangle with the help of elementary functions.

In order to determine parameters of the non-elementary function defined by the Schwarz-Christoffel integral let us place the vertices of the triangle into the points $w_1 = 0$ and $w_2 = h$. Since $w(0) = 0$ the constant C_2 in the expression (1.12.36) vanishes.

Transforming the expression $C_1 (\xi - 1)^{\beta-1}$ into $C(1 - \xi)^{\beta-1}$ (where $C = C_1 e^{i\pi(\beta-1)}$) represent the mapping function as

$$w = C \int_0^z \xi^{\alpha-1} (1 - \xi)^{\beta-1} \, d\xi. \qquad (1.12.37)$$

The function $B_z(\alpha, \beta) = \int_0^z \xi^{\alpha-1} (1 - \xi)^{\beta-1} \, d\xi$ is called *non-complete beta-function*.[99, 174, 175] It is a non-elementary function analytic in the half-plane Im $z > 0$. Substituting $z = 1$ into the equality (1.12.37) and taking into account that $w(1) = h$ we can find the value of the constant C:

$$C = \frac{h}{B(\alpha, \beta)},$$

where $B(\alpha, \beta) = B_1(\alpha, \beta) = \int_0^1 \xi^{\alpha-1} (1 - \xi)^{\beta-1} \, d\xi$ is the ordinary beta-function.

Angular domains

In the triangular angular domain the integer angle can be 2π, 0, $-\pi$, -2π. Accordingly there exist four types of angular domains with one integer angle, a half-plane to be mapped by elementary functions.

a) The integer angle equal to 2π. In this case the angles of the considered domain are 2π, $\pi\theta$, $-\pi - \pi\theta$, where $0 < \theta < 1$. The domain is the

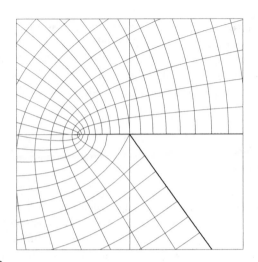

FIGURE 1.86

angle with the star-shaped cut, for which the mapping function was built in the Subsection 1.12.1:

$$w = Cz^{\theta}(z-1).\qquad(1.12.38)$$

The mapping of the Cartesian net in the half-plane $\mathrm{Im}\, z > 0$ carried out by the function (1.12.38) with $\theta = 0.7$ is presented in Fig. 1.86.

b) The integer angle equal to 0. Suppose

$$a_1 = -1, \qquad a_2 = 1, \qquad a_3 = \infty;$$
$$\alpha = \theta, \qquad \beta = 1 - \theta, \qquad \gamma = 0,$$

where θ is a rational number, $0 < \theta < 1$.

If $w(0) = 0$ the integral (1.12.1)–(1.12.4) yields

$$w = C_1 \int_0^z (\xi+1)^{\theta-1}(\xi-1)^{-\theta}\, d\xi.\qquad(1.12.39)$$

When $\theta = p/q$ is a rational number, the integral 1.12.39 is reduced by the change of the integration variable to the integral of the rational function and, therefore, is expressed as elementary functions. For example, this can be done with the help of the substitution

$$t = \left(\frac{\xi-1}{\xi+1}\right)^{1/q},\qquad(1.12.40)$$

i.e.,

$$\xi = \frac{1+t^q}{1-t^q}.$$

Example 1.29
When $\theta = 1/2$ the integral (1.12.39) can be written as

$$w = iC_1 \int_0^z \frac{d\xi}{\sqrt{1 - \xi^2}} = iC_1 \arcsin z \, .$$

This elementary function maps a half-plane onto a half-strip.

Example 1.30
When $\theta = 3/4$ the integral (1.12.39) yields

$$w = C_1 \int_0^z \left(\frac{\xi - 1}{\xi + 1} \right)^{1/4} \frac{d\xi}{\xi - 1} \, .$$

After the change of the variable (1.12.40) $\xi = (1 + t^4)/(1 - t^4)$ it is reduced to

$$w = 4C_1 \int \frac{dt}{1 - t^4} = 2C_1 \left(\arctan t - \operatorname{arth} t \right) .$$

The mapping of the Cartesian net in the upper half-plane realized by the function

$$t = \left(\frac{z - 1}{z + 1} \right)^{1/4} \, ,$$

$$w = 2 \arctan t + 2 \operatorname{arth} t - i \frac{\pi}{2} - \pi \, ,$$

is presented in Fig. 1.87.

c) The integer angle equal to $-\pi$. Suppose

$$a_1 = -1, \qquad a_2 = 1, \qquad a_3 = \infty \, ;$$
$$\alpha = 1 - \theta, \qquad \beta = 1 + \theta, \qquad \gamma = -1 \, ,$$

where θ is a rational number, $-1 < \theta < 1$, $\theta \neq 0$. When $w(1) = 0$ the Schwarz-Christoffel integral yields

$$w = C_1 \int_1^z \left(\frac{\xi - 1}{\xi + 1} \right)^{\theta} d\xi \, . \tag{1.12.41}$$

If $\theta = p/q$, where p and q are integer numbers, this integral can be rationalized after the variable change (1.12.40).

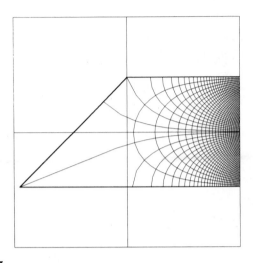

FIGURE 1.87

Example 1.31

When $\theta = 1/2$ the integral (1.12.41) can be represented as

$$w = C_1 \int_1^z \frac{1 - \xi}{\sqrt{1 - \xi^2}} \, d\xi = iC_1 \left(\arcsin z + \sqrt{1 - z^2} - \frac{\pi}{2} \right).$$

The mapping of the Cartesian net in the upper half-plane realized by the function with $C_1 = 1$ is presented in Fig. 1.88.

c) The integer angle equal to -2π. Suppose

$$a_1 = -1, \qquad a_2 = 1, \qquad a_3 = \infty;$$
$$\alpha = 1 + \theta, \qquad \beta = 2 - \theta, \qquad \gamma = -2,$$

where θ is a rational number, $0 < \theta < 1$. The Schwarz-Christoffel integral yields

$$w = C_1 \int_{z_0}^z (\xi + 1)^\theta \, (\xi - 1)^{1-\theta} \, d\xi + C_2. \qquad (1.12.42)$$

This integral can be rationalized after the variable change (1.12.40).

Example 1.32

When $\theta = 1/2$ and $w(0) = 0$ the integral (1.12.42) can be represented as

$$w = iC_1 \int_0^z \sqrt{1 - \xi^2} \, d\xi = C_1 \frac{i}{2} \left(z\sqrt{1 - z^2} + \arcsin z \right). \qquad (1.12.43)$$

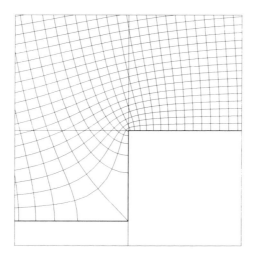

FIGURE 1.88

The mapping of the Cartesian net in the upper half-plane realized by the function with $C_1 = 2$ is presented in Fig. 1.89.

Triangular strips

The strip $0 < \operatorname{Im} z < \pi$ is the adequate canonical domain for curvilinear strips. In order to construct the conformal mapping of the canonical domain onto the considered triangular strip it is necessary to map the canonical strip onto the half-plane $\operatorname{Im} z_1 > 0$ with the help of the function $z_1 = e^z$ and then with the help of the Schwarz-Christoffel integral to map the half-plane onto the considered triangular strip, so that the points $z_1 = 0$ and $z_1 = \infty$ would be mapped into the two-folded boundary point $w = \infty$.

An integer angle in the triangular strip can be equal either to 2π or to 0.[‡] Accordingly there exist two types of triangular strips, onto which it is possible to map the canonical domain by elementary functions.

a) The integer angle equal to 2π. Let the angles of the triangular strip be 2π, $-\pi\theta$, $-\pi + \pi\theta$, where $0 < \theta < 1$. This domain represents the half-plane with a cut along a ray. It can be considered as a star-shaped domain.

[‡]The angle is possible to equal $-\pi$ in the triangular strip in combination with the angles 0 and 2π only, i.e., in the case when all the angles of the triangular domain are integer. This case was considered above.

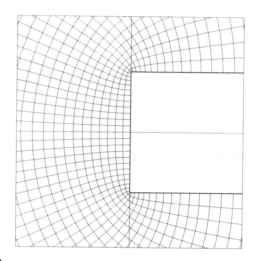

FIGURE 1.89

In accordance with the formulae of Subsection (1.12.1) the function, which maps the half-plane Im $z_1 > 0$ onto the considered star-shaped strip, takes the form

$$w = C \, z_1^{-\theta}(z_1 - 1) \,. \tag{1.12.44}$$

To map the strip $0 < \text{Im } z < \pi$ onto the considered triangular strip it is necessary to make a change of the variable $z_1 = e^z$ in the function (1.12.44). The mapping of the Cartesian net inside the strip $0 < \text{Im } z < \pi$ realized by the function (1.12.44) with $\theta = 0.43$, $z_1 = e^z$ is presented in Fig. 1.90.

b) The integer angle equal to 0.   Suppose

$$a_1 = -1, \quad a_2 = 0, \quad a_3 = \infty \,;$$
$$\alpha = 1 + \theta, \quad \beta = 0, \quad \gamma = -\theta \,,$$

where $\theta = p/q$ is a rational number from the interval $0 < \theta < 1$. Under the condition $w(-1) = 0$ the Schwarz-Christoffel integral yields

$$w = C_1 \int_{-1}^{z_1} (\xi + 1)^\theta \, \frac{d\xi}{\xi} \,. \tag{1.12.45}$$

When $\theta = p/q$ the integral (1.12.45) is rationalized by the substitution $\xi = t^q - 1$.

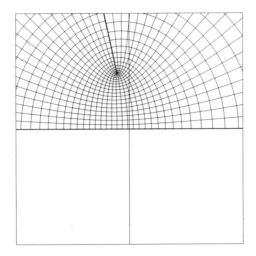

FIGURE 1.90

Example 1.33

When $\theta = 1/2$ the integral (1.12.45) is equal to

$$w = C_1 \left(2t - \ln \frac{1+t}{1-t} \right) = 2C_1 \left(t - \operatorname{arth} t \right), \qquad (1.12.46)$$

where $t = \sqrt{z_1 + 1} = \sqrt{e^z + 1}$. The mapping of the strip $0 < \operatorname{Im} z < \pi$ realized by the function (1.12.46) with $C_1 = 1$ is presented in Fig. 1.91.

_____

Exteriors of triangles

Let D be the triangular domain of the w-plane being the exterior of the triangle with the vertices A_1, A_2, A_3. Let the angles of the domain D at the vertices (i.e., the exterior angles of the triangle) be equal to $\pi\alpha$, $\pi\beta$, $\pi\gamma$, respectively. These angles satisfy the conditions $1 < \alpha < 2$, $1 < \beta < 2$, $1 < \gamma < 2$, $\alpha + \beta + \gamma = 5$. The adequate canonical domain for the domain D is the exterior of the unit circle. Let the function $z = F(w)$ realize the conformal mapping of the domain onto the exterior of the circle $|z| > 1$, the point $w = \infty$ being mapped into the point $z = \infty$ and the points A_1, A_2, A_3 being mapped into some points a_1, a_2, a_3 of the unit circumference, correspondingly. The function $f = F^{-1}$, which realizes the inverse mapping of the exterior of the circle onto that of the triangle, is in accordance with

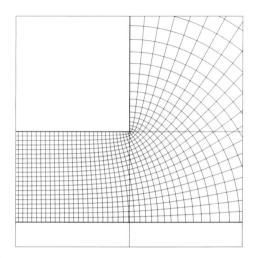

FIGURE 1.91

the formula (1.12.5):

$$w = C_1 \int_{z_0}^{z} (\xi - a_1)^{\alpha-1} (\xi - a_2)^{\beta-1} (\xi - a_3)^{\gamma-1} \frac{d\xi}{\xi^2} + C_2 \,.$$

This function is not elementary for any permissible values of the parameters α, β, γ.

Exercises

1. When $\theta = 1/4$, $C_1 = 1$ the integral (1.12.41) is equal to

$$w = \frac{4t}{1 - t^2} - 2\arctan t - 2\operatorname{arth} t, \qquad (1.12.47)$$

   where $t = ((z-1)/(z+1))^{1/4}$. Create the mapping of the Cartesian net in the upper half-plane $\operatorname{Im} z > 0$ carried out by the function (1.12.47).

2. When $\theta = 1/4$, $C_1 = 1/2$ the integral (1.12.42) is equal to

$$w = \frac{3t + t^5}{(1 - t^2)^2} - \frac{3}{2}\left(\arctan t + \operatorname{arth} t - i\frac{\pi}{4}\right), \qquad (1.12.48)$$

   where $t = ((z-1)/(z+1))^{1/4}$. Create the mapping of the Cartesian net in the upper half-plane $\operatorname{Im} z > 0$ carried out by the function (1.12.48).

3. When $\theta = 1/4$, $C_1 = 1/2$ the integral (1.12.45) is equal to

$$w = 2t - \arctan t - \operatorname{arth} t. \qquad (1.12.49)$$

   Create the mapping of the Cartesian net in the strip $0 < \operatorname{Im} z < \pi$ carried out by the function (1.12.49) with $t = (e^z + 1)^{1/4}$.

4. When $\theta = 3/4$, $C_1 = 1/2$ the integral (1.12.45) is equal to

$$w = \frac{2}{3} t^3 - \operatorname{arth} t + \arctan t . \qquad (1.12.50)$$

Create the mapping of the Cartesian net in the strip $0 < \operatorname{Im} z < \pi$ realized by the function (1.12.50) with $t = (e^z + 1)^{1/4}$.

1.12.4   Mappings of quadrangular domains

In this Subsection we will use a more brief term "quadrangle $(\alpha, \beta, \gamma, \delta)$" instead of "quadrangular domain with the angles $\pi\alpha$, $\pi\beta$, $\pi\gamma$, $\pi\delta$". In the case of a finite or infinite quadrangle the numbers α, β, γ, δ satisfy the relation $\alpha + \beta + \gamma + \delta = 2$, in the case of the exterior of a quadrangle — the relation $\alpha + \beta + \gamma + \delta = 6$. It should be mentioned that the appearance of a quadrangle depends not only on the values of angles but on their sequence. The shape of a quadrangle does not change under a cyclic permutation of the numbers $\alpha, \beta, \gamma, \delta$. A permutation of these numbers in the inverse order corresponds to the mirror reflection of the quadrangle.

The *permutations* of the numbers $\alpha, \beta, \gamma, \delta$ other than cyclic and (or) inverse order permutations are called *non-trivial* ones. They define non-trivial transformations of the quadrangular domain.

A conformal mapping of a half-plane onto the given finite or infinite quadrangle is carried out by the Schwarz-Christoffel formula (1.12.1-1.12.4) with the real parameters a_1, a_2, a_3, a_4 unknown in advance. Three of them can be arbitrarily assigned, but the fourth one should be defined. Therefore passing from triangles to quadrangles we first meet the problem of determination of the parameters in the Schwarz-Christoffel integral. This problem has no analytic solution. Instead of the problem of the determination of the parameters to obtain the necessary quadrangle we will investigate the dependence of the Schwarz-Christoffel integral on the parameters. In the program CONFORM the parameters of the Schwarz-Christoffel integral are chosen experimentally in order to obtain the required shape of the domain.

Further, the question arises on the determination of the function which maps a half-plane onto a quadrangle. In some cases the Schwarz-Christoffel integral is expressed as elementary functions. First of all it concerns the quadrangles with four integer angles, the integrand function in the Schwarz-Christoffel integral is rational. In the cases when the quadrangle has two integer and two rational angles, it is possible to reduce the Schwarz-Christoffel integral to the integral of the rational function with the help of the variable change.

Naturally the question arises whether it is possible to express the Schwarz-Christoffel integral as elementary functions in the case of the quadrangle with one integer angle. In the general case the answer to the question is negative. Indeed, let an integer angle of the quadrangle be equal to 2π

(choose, for example, $\delta = 2$). Assume

$$a_1 = 0, \quad a_2 = 1, \quad a_3 = \infty, \quad a_4 = -p;$$
$$\alpha_1 = \alpha, \quad \alpha_2 = \beta, \quad \alpha_3 = \gamma, \quad \alpha_4 = 2.$$

Schwarz-Christoffel integral for this quadrangle having the form

$$w = C_1 \int_{z_0}^{z} \xi^{\alpha-1}(\xi - 1)^{\beta-1}(\xi + p)\, d\xi + C_2, \tag{1.12.51}$$

represents the sum of two differential binomials. In accordance with the Chebyshev's theorem the integral on the differential binomials is expressed as an elementary function only in the cases when either α, or β, or $\alpha + \beta$ is an integer number. Applying this to the quadrangle it means that at least one of the numbers α, β, γ should be an integer, i.e., besides the integer angle $\delta = 2$ the quadrangle should have one more integer angle. Thus if the quadrangle has one integer angle $\delta = 2$, each term in the integral (1.12.51) is a non-elementary function.

Nevertheless the presence of two integer angles is not the necessary condition for the whole integral (1.12.51) to be expressed as elementary functions. It can be demonstrated by the following example.

Example 1.34
An elementary function

$$w = z^{0.5}(z - 1)^{0.9}$$

maps the half-plane Im $z > 0$ onto the star-shaped quadrangular domain without integer angles except for the angle 2π (Fig. 1.92).

Thus in the case of the quadrangle with the single integer angle the Schwarz-Christoffel integral in general is a non-elementary function, but in some exceptional cases it expresses as elementary functions. In the case of the quadrangle without integer angles (in particular, for the finite quadrangle) the Schwarz-Christoffel integral is not expressed as elementary function. For the rectangles (and other quadrangles with four right angles) the Schwarz-Christoffel integral is expressed as non-elementary functions — the elliptical integrals. Such conformal mappings are considered in the next Section 1.13.

In this Subsection the following classes of quadrangles are considered, for which that the Schwarz-Christoffel integral is expressed as elementary functions:

  I. Quadrangles with four integer angles.

  II. Quadrangles with two integer and two rational angles.

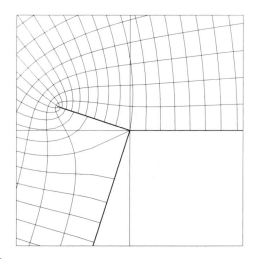

FIGURE 1.92

III. Some quadrangles with one integer angle.

For the quadrangles of 1 and 2 classes it is possible to consider all the kinds and to give their classification. For the quadrangles of class 3 we can not affirm that besides the considered quadrangles there do not exist others with one integer angle, for which the Schwarz-Christoffel integral is expressed as elementary functions.

I. Quadrangles with four integer angles

There exist three types of the quadrangles with four integer angles: $(2, 0, 2, -2)$, $(2, -1, 2, -1)$ and $(0, 0, 0, 2)$. The first two types represent the plane with two cuts along rays of parallel straight lines, the third type represents the rectilinear strip with a cut along a ray. Mappings of such domains were considered in the Subsection 1.12.2 (Examples 1.22–1.24, 1.28).

II. Quadrangles with two integer angles

All the possible quadrangles with two integer angles can be built by adding two non-integer angles to the pair of the integer ones in such a way that the sum of all the angles would be equal to 2π. Then various non-trivial permutations of four angles should be applied.

Possible integer values for the numbers α, β, γ, δ are 2, 0, -1, -2. Ten different pairs can be composed from these numbers:

$$(2,2), \quad (2,0), \quad (2,-1), \quad (2,-2), \quad (0,0),$$
$$(0,-1), \quad (0,-2), \quad (-1,-1), \quad (-1,-2), \quad (-2,-2).$$

The two latest combinations are impossible because the sum of angles at the point at infinity can not exceed 2. Combinations of angles $(0,-2)$ and $(-1,-1)$ are possible only when all the angles of the quadrangle are the integer ones. These cases have already been considered above. Thus 6 mutual combinations of integer angles remain, that will be considered separately for quadrangular angular domains, quadrangular strips and exteriors of quadrangles.

A. Quadrangular angular domains. Only two combinations of integer angles are possible for quadrangular angular domains: $(2,-1)$ and $(2,-2)$.

1. Integer angles $(2,-1)$. For integer angles 2 and -1 there exist quadrangles $(2,-1,\theta,1-\theta)$ and $(2,\theta,-1,1-\theta)$. Assume in the formulae (1.12.1–1.12.4)

$$a_1 = p, \quad a_2 = \infty, \quad a_3 = -1, \quad a_4 = 1;$$
$$\alpha = 2, \quad \beta = -1, \quad \gamma = \theta \quad \delta = 1 - \theta;$$
$$w(1) = 0,$$

where $p > 0$, $0 < \theta < 1$. The Schwarz-Christoffel integral takes the form

$$w = C_1 \int\limits_1^z (\xi - p)(\xi + 1)^{\theta - 1}(\xi - 1)^{-\theta} d\xi + C_2, \qquad (1.12.52)$$

and if $\theta = m/n$ is a rational number, then the integral (1.12.52) is rationalized by the substitution

$$t = \left(\frac{\xi - 1}{\xi + 1}\right)^{1/n}, \quad \xi = \frac{1 + t^n}{1 - t^n}.$$

Example 1.35
When $\theta = 1/2$ the integral (1.12.52) is calculated without rationalization:

$$w = -i\,C_1 \int_1^z \frac{\xi - p}{\sqrt{1 - \xi^2}} d\xi = i\,C_1 \left[\sqrt{1 - z^2} + p\arcsin z - p\frac{\pi}{2}\right]. \qquad (1.12.53)$$

The mapping of the Cartesian net in the half-plane $\operatorname{Im} z > 0$ carried out by the function (1.12.53) with $p = 4$ is presented in Fig. 1.93. The mapping is

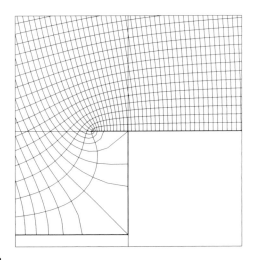

FIGURE 1.93

a step with the horizontal projection. If to take $0 < p < 1$ the step with the vertical projection would be obtained. This case is presented in Catalog 3 (Domain 18). When $p = 1$ the projection disappears and the domain turns into the triangle considered in Subsection 1.12.3.

2. Integer angles $(2, -2)$. Two types of quadrangular angular domains exist with integer angles 2 and -2: $(2, -2, 2 - \theta, \theta)$ and $(2, \theta, -2, 2 - \theta)$. Assume in the formulae (1.12.1)–(1.12.4)

$$a_1 = p, \quad a_2 = \infty, \quad a_3 = -1, \quad a_4 = 1;$$
$$\alpha = 2, \quad \beta = -2, \quad \gamma = 2 - \theta, \quad \delta = \theta,$$

where $p > 0$, $0 < \theta < 1$. The Schwarz-Christoffel integral takes the form

$$w = C_1 \int_0^z (\xi - p)(\xi + 1)^{1-\theta}(\xi - 1)^{\theta-1} d\xi + C_2 . \qquad (1.12.54)$$

When $\theta = m/n$ the integral is rationalized by the substitution

$$t = \left(\frac{\xi - 1}{\xi + 1} \right)^{1/n}, \quad \xi = \frac{1 + t^n}{1 - t^n} .$$

Example 1.36
When $\theta = 1/2$ the integral (1.12.54) is calculated without rationalization:

$$w = -i\, C_1 \int_0^z \frac{(\xi - p)(\xi + 1)}{\sqrt{1 - \xi^2}} d\xi + C_2$$

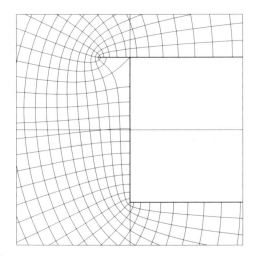

FIGURE 1.94

$$= i\,C_1 \left[\left(1 - p + \frac{z}{2}\right)\sqrt{1 - z^2} + \left(p - \frac{1}{2}\right)\arcsin z \right] + \overline{C_2} \quad (1.12.55)$$

where $\overline{C_2} = C_2 - iC_1(p-1)$. The mapping of the Cartesian net of the upper half-plane carried out by the function (1.12.55) with $p = 1.5$, $C_1 = 1/(p - 1/2)$, $\overline{C_2} = 0$ is presented in Fig. 1.94.

The image of the upper half-plane in the mapping (1.12.55) is the exterior of a half-strip with a cut along some segment. When $p > 1$ the cut is directed horizontally, when $1/2 \leq p < 1$ — vertically. When $p = 1$ the cut is absent and the function (1.12.55) coincide with (1.12.43).

When $p = 1/2$ the boundary of the domain is a star-shaped cut. When $p < 1/2$ the function (1.12.55) is not univalent in the upper half-plane.

B. Quadrangular strips. Quadrangular strips are obtained under the following combinations of integer angles: $(2,2)$, $(2,0)$, $(2,-1)$, $(0,0)$ and $(0,-1)$.

3. Integer angles $(2,2)$. Angles $(2,2)$ are the angles of the quadrangle $(2, -1 - \theta, 2, \theta - 1)$, which represent the plane with two cuts along rays of non-parallel straight lines (we suppose θ to be an arbitrary number from the interval $0 < \theta < 1$). The cases when the rays belong to parallel straight lines were considered above in Subsection 1.12.2.

Assume in the formulae (1.12.1)–(1.12.4)

$$a_1 = 0, \qquad a_2 = 1, \qquad a_3 = \infty, \qquad a_4 = -p;$$
$$\alpha = -1 + \theta, \quad \beta = 2, \quad \gamma = -1 - \theta, \quad \delta = 2,$$

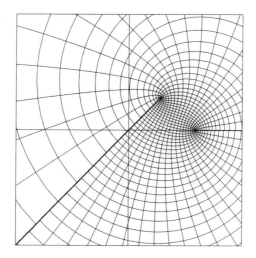

FIGURE 1.95

where $p > 0$. The Schwarz-Christoffel integral takes the form

$$w = C_1 \int_{z_0}^{z_1} \xi^{\theta-2}(\xi - 1)(\xi + p)\, d\xi + C_2$$

$$= C_1 z_1^{\theta-1} \left(\frac{z_1^2}{\theta + 1} + z_1 \frac{p-1}{\theta} - \frac{p}{\theta - 1} \right) + C_2. \qquad (1.12.56)$$

Example 1.37
The mapping of the strip $0 < \operatorname{Im} z < \pi$ carried out by the function (1.12.56) with $z_1 = e^z$, $\theta = 0.25$, $p = 5$, $C_1 = 1$, $C_2 = 0$ is presented in Fig. 1.95.

4. Integer angles $(2,0)$. Quadrangular strips of three types can be built with the integer angles $(2,0)$: $(2,0,\theta,-\theta)$, $(2,\theta,0,-\theta)$ and $(2,-\theta,\theta,0)$. In the two first cases it is assumed $0 < \theta < 1$ while in the third case $0 < \theta < 2$.

Substitute into the formulae $(1.12.1)$–$(1.12.4)$

$$
\begin{array}{llll}
a_1 = p, & a_2 = \infty, & a_3 = -1, & a_4 = 0; \\
\alpha = 2, & \beta = -\theta, & \gamma = \theta & \delta = 0; \\
w(-1) = 0, & & &
\end{array}
$$

assuming $p \neq 0$, $p \neq 1$. The Schwarz-Christoffel integral takes the form

$$w = C_1 \int_{-1}^{z_1} \xi^{-1}(\xi + 1)^{\theta-1}(\xi - p)\, d\xi. \qquad (1.12.57)$$

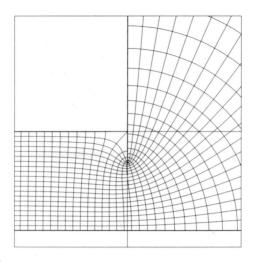

FIGURE 1.96

If $\theta = m/n$ is a rational number, the integral is reduced by the variable change $\xi = t^n - 1$ to

$$w = C_1 n \int_0^t \left(1 - \frac{p}{t^n - 1}\right) t^{m-1} \, dt. \qquad (1.12.58)$$

The function (1.12.58) (where $t = (z_1 + 1)^{1/n}$) maps the half-plane $\operatorname{Im} z > 0$ onto a quadrangle, with the appearance depending on the value of p parameter. The quadrangle $(2, 0, -\theta, \theta)$ is obtained for $0 < p < 1$, the quadrangle $(2, \theta, 0, -\theta)$ — for $p > 1$, the quadrangle $(2, -\theta, \theta, 0)$ — for $p < 0$.

Example 1.38

When $\theta = 1/2$ ($m = 1$, $n = 2$) the integral (1.12.57) takes the form

$$w = 2C_1 (t - p \operatorname{arth} t). \qquad (1.12.59)$$

The mapping of the Cartesian net in the strip $0 < \operatorname{Im} z < \pi$ carried out by the function (1.12.59) with $z_1 = e^z$, $t = \sqrt{z_1 + 1}$, $p = 3$ is presented in Fig. 1.96. The mappings carried out by this function with other values of the parameter p are presented in Catalog 4, Domains 27, 28.

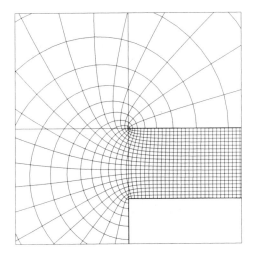

FIGURE 1.97

Example 1.39

When $\theta = 3/2$ the integral (1.12.57) takes the form after the change $\xi = t^2 - 1$:

$$w = 2C_1 \int_0^t \left(t^2 - p + \frac{p}{1 - t^2}\right) dt = 2C_1 \left(\frac{t^3}{3} - pt + p\,\mathrm{arth}\,\frac{1}{t}\right). \quad (1.12.60)$$

The mapping of the Cartesian net in the strip $0 < \mathrm{Im}\, z < \pi$ carried out by the function (1.12.60) with $z_1 = e^z$, $t = \sqrt{z_1 + 1} = \sqrt{e^z + 1}$, $p = 1.55$, $C = 1/2p$ is presented in Fig. 1.97.

5. Integer angles $(2, -1)$. The quadrangular strip with integer angles $(2, -1)$ is the quadrangle $(-1, 2, -\theta, 1 + \theta)$, where $0 < \theta < 1$. In this case assume

$$a_1 = 0, \quad a_2 = p, \quad a_3 = \infty, \quad a_4 = -1;$$
$$\alpha = -1, \quad \beta = 2, \quad \gamma = -\theta, \quad \delta = 1 + \theta;$$
$$w(-1) = 0, \quad p > 0.$$

The Schwarz-Christoffel integral yields

$$w = C_1 \int_{-1}^{z_1} (\xi + 1)^\theta (\xi - p) \frac{d\xi}{\xi^2}. \quad (1.12.61)$$

If $\theta = m/n$ is a rational number, the integral is rationalized by the substitution $\xi = t^n - 1$.

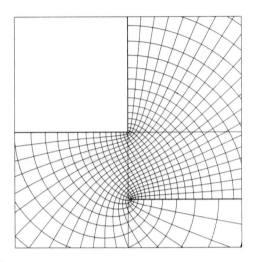

FIGURE 1.98

Example 1.40

When $\theta = 1/2$ the change $\xi = t^2 - 1$ reduces the integral to

$$w = 2C \int_0^t \frac{t^2(t^2 - p - 1)}{(t^2 - 1)^2} \, dt = C \left[2t + \frac{pt}{t^2 - 1} - (2 - p) \operatorname{arth} t \right].$$

The mapping of the strip $0 < \operatorname{Im} z < \pi$ by this function with $t = \sqrt{z_1 + 1} = \sqrt{e^z + 1}$, $p = 0.15$, $C = -1/2$ is presented in Fig. 1.98.

6. Integer angles $(0,0)$. Two types of quadrangular strips are possible for the integer angles $(0,0)$: $(0, 0, 1 - \theta, 1 + \theta)$ and $(0, 1 - \theta, 0, 1 + \theta)$. For these quadrangles assume

$$\begin{array}{cccc} a_1 = 0, & a_2 = p, & a_3 = 1, & a_4 = \infty; \\ \alpha = 0, & \beta = 1 - \theta, & \gamma = 1 + \theta, & \delta = 0; \\ w(1) = 0. \end{array}$$

The quadrangle $(0, 1 - \theta, 1 + \theta, 0)$ is obtained for $0 < p < 1$ and the quadrangle $(0, 1 - \theta, 0, 1 + \theta)$ — for $p < 0$.

The Schwarz-Christoffel integral yields

$$w = C_1 \int_1^{z_1} \xi^{-1} (\xi - p)^{-\theta} (\xi - 1)^\theta d\xi. \tag{1.12.62}$$

If $\theta = m/n$ is a rational number the integral (1.12.62) is reduced by the change $t = [(\xi - 1)/(\xi - p)]^{1/n}$, $\xi = (1 - pt^n)/(1 - t^n)$ to

$$w = C_1 n (1 - p) \int_0^t \frac{t^{n+m-1} dt}{(1 - t^n)(1 - pt^n)}$$

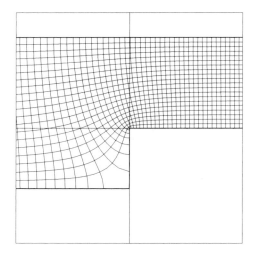

FIGURE 1.99

$$= C_1 n \int_0^t \left(\frac{1}{1 - t^n} - \frac{1}{1 - p t^n} \right) t^{m-1} dt . \qquad (1.12.63)$$

Example 1.41

When $\theta = 1/2$ $(n = 2,\ m = 1)$, $p = q^2 > 0$ the integral (1.12.63) is equal to

$$w = 2C_1 \left(\operatorname{arth} t - \frac{1}{q} \operatorname{arth} qt \right). \qquad (1.12.64)$$

The mapping of the Cartesian net in the strip $0 < \operatorname{Im} z < \pi$ realized by the function (1.12.64) with $z_1 = e^z$, $t = \sqrt{(z_1 - 1)/(z_1 - q^2)}$, $q = 0.6$, $C_1 = 1/2$ is presented in Fig. 1.99.

Example 1.42

When $\theta = 1/2$, $p = -q^2 < 0$ the integral (1.12.63) is expressed slightly differently

$$w = A \left(q \operatorname{arth} t - \arctan qt \right), \qquad (1.12.65)$$

where $A = 2C_1/q$. The mapping of the Cartesian net in the strip $0 < \operatorname{Im} z < \pi$ realized by the function (1.12.65) with $z_1 = e^z$, $t = \sqrt{(z_1 - 1)/(z_1 + q^2)}$, $q = 0.66$, $A = 1$ is presented in Fig. 1.100.

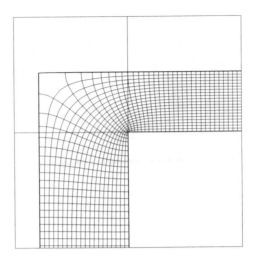

FIGURE 1.100

7. Integer angles $(0, -1)$. Two types of quadrangular strips with integer angles 0 and -1 exist: $(-1, 0, 2 - \theta, 1 + \theta)$ and $(-1, 2 - \theta, 0, 1 + \theta)$. For such quadrangles assume

$$a_1 = \infty, \quad a_2 = 0, \quad a_3 = p, \quad a_4 = 1;$$
$$\alpha = -1, \quad \beta = 0, \quad \gamma = 2 - \theta, \quad \delta = 1 + \theta.$$

The quadrangle $(-1, 0, 2-\theta, 1+\theta)$ is obtained for $0 < p < 1$, the quadrangle $(-1, 2 - \theta, 0, 1 + \theta)$ — for $p < 0$.

The Schwarz-Christoffel integral for such quadrangles yields

$$w = C_1 \int_1^z \xi^{-1}(\xi - p)^{1-\theta}(\xi - 1)^\theta d\xi + C_2 . \tag{1.12.66}$$

If $\theta = m/n$ is the rational number, the integral (1.12.66) can be rationalized by the change $t = [(\xi - 1)/(\xi - p)]^{1/n}$, $\xi = (1 - pt^n)/(1 - t^n)$:

$$w = C_1(1 - p) \int_0^t \left[\frac{nt^n}{(1 - t^n)^2} - \frac{npt^n}{(1 - t^n)(1 - pt^n)} \right] t^{m-1} dt + C_2 . \tag{1.12.67}$$

When $m = 1$ the integral is expressed especially simply:

$$w = C_1 \left[\frac{1 - p}{1 - t^n} t + (1 + pn - p) \int \frac{dt}{1 - t^n} + np \int \frac{dt}{1 - pt^n} \right] + C_2 . \tag{1.12.68}$$

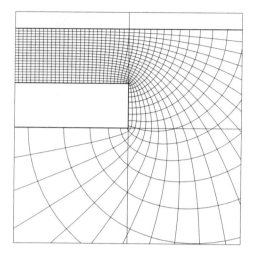

FIGURE 1.101

Example 1.43
When $\theta = 1/2$, $p = q^2 > 0$ the integral (1.12.68) is equal to

$$w = C_1 \left[\frac{1-q^2}{1-t^2} t - (1+q^2) \operatorname{arth} t + 2q \operatorname{arth} qt \right] + C_2 . \qquad (1.12.69)$$

The mapping of the Cartesian net in the strip $0 < \operatorname{Im} z < \pi$ realized by the function (1.12.69) with $z_1 = e^z$, $t = \sqrt{(z_1 - 1)/(z_1 - q^2)}$, $q = 0.3$, $C_1 = -1$, $C_2 = 0$ is presented in Fig. 1.101.

Example 1.44
When $\theta = 1/2$, $p = -q^2 < 0$ the integral (1.12.68) is expressed somewhat differently

$$w = C_1 \left[\frac{1+q^2}{1-t^2} t - (1-q^2) \operatorname{arth} t - 2q \arctan qt \right] + C_2 . \qquad (1.12.70)$$

The mapping of the Cartesian net in the strip $0 < \operatorname{Im} z < \pi$ realized by the function (1.12.70) with $z_1 = e^z$, $t = \sqrt{(z_1 - 1)/(z_1 + q^2)}$, $q = 0.8$, $C_1 = 1/q$, $C_2 = \pi/2$ is presented in Fig. 1.102.

C. Exteriors of quadrangles. Exteriors of finite quadrangles can not be obtained by the conformal mapping of a canonical domain (the exterior of a circle) with the help of elementary functions, except for the case of a star-shaped quadrangle.

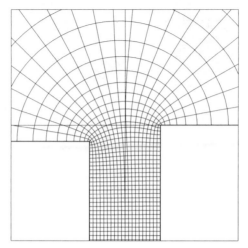

FIGURE 1.102

8. Quadrangle $(1-\theta, 2, 1+\theta, 2)$. Let an infinite quadrangular domain be an exterior of a cut being a two-linked broken line with the angles $(1-\theta)$ and $(1+\theta)$ in the origin. Assume in the formula (1.12.12)

$$a_1 = 1, \qquad a_2 = e^{ip};$$
$$\alpha_1 = 1 - \theta, \quad \alpha_2 = 1 + \theta.$$

The function which maps the exterior of the unit circle onto the considered star-shaped polygon has the form

$$w = C\frac{(z-1)^{1-\theta}(z-e^{ip})^{1+\theta}}{z} . \qquad (1.12.71)$$

In order to orient one of the links of the broken line along the real axis we should assume

$$C = A\exp\left[i\left(\pi\theta - \frac{p}{2} - \frac{p\theta}{2}\right)\right], \qquad (1.12.72)$$

where $A > 0$ is a scaling factor. To extract the principal branch of the multi-valued function (1.12.71) continuous outside the circle $|z| > 1$ it should be represented in the form

$$w = Cz\left(1 - \frac{1}{z}\right)^{1-\theta}\left(1 - \frac{e^{ip}}{z}\right)^{1+\theta} . \qquad (1.12.73)$$

The mapping of the polar net in the exterior of the unit circle realized by the function (1.12.72)–(1.12.73) with $\theta = 0.25$, $p = 4$ is presented in Fig. 1.103. Varying the value of the parameter p $(0 < p < 2\pi)$ in the

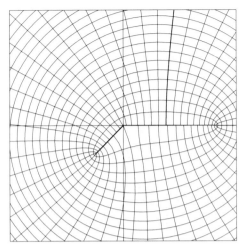

FIGURE 1.103

formulae (1.12.72)–(1.12.73) makes it possible to change the ratio of lengths of broken line links. The lengths of the links are equal when $p = \pi$.

III. Quadrangles with one integer angle

Some types of quadrangles with one integer angle can be obtained as conformal mappings of canonical domains with the help of elementary functions. We consider star-shaped quadrangles and symmetrical quadrangles with one integer angle. Apparently these cases (along with those considered above) exhaust the class of quadrangular domains mapped by elementary functions.

A. The star-shaped quadrangle $(\alpha, 2, \beta, -\alpha - \beta)$ represents an angular star-shaped domain. The function mapping the half-plane Im $z > 0$ onto the considered quadrangle in accordance with (1.12.6) is

$$w = C\, z^{\alpha} (z - 1)^{\beta} . \tag{1.12.74}$$

An example of the mapping realized by the function (1.12.74) is presented in Fig. 1.92.

B. The star-shaped quadrangle $(-\alpha, 2, -\beta, \alpha + \beta)$ is a star-shaped strip. The function mapping the half-plane Im $z > 0$ onto the considered domain

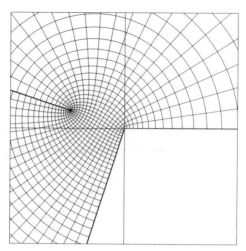

FIGURE 1.104

in accordance with (1.12.10) is

$$w = C\, z^{-\alpha}(z-1)^{\alpha+\beta}. \tag{1.12.75}$$

An example of the mapping of the Cartesian net in the strip $0 < \operatorname{Im} \xi < \pi$ realized by the function (1.12.75) with $\alpha = 0.5$, $\beta = 0.9$, $z = e^\xi$ is presented in Fig. 1.104.

C. Symmetrical quadrangular angular domain with the integer angle 0. It is possible to use the symmetry principle to construct the mapping of the symmetrical quadrangle $(0, 1 - \theta, 2\theta, 1 - \theta)$. Assume in the formulae (1.12.1)–(1.12.4)

$$\begin{aligned}
&a_1 = \infty, \quad a_2 = -1, \quad a_3 = 0, \quad a_4 = p; \\
&\alpha = 0, \quad \beta = 1 - \theta, \quad \gamma = 2\theta, \quad \delta = 1 - \theta.
\end{aligned}$$

Since the considered domain is symmetrical with respect to the bisectrix of the angle $\gamma = 2\theta$, the points $a_2 = -1$ and $a_4 = p$ should be symmetrical with respect to the point $a_3 = 0$, i.e., it is necessary $a_4 = p = 1$. The Schwarz-Christoffel integral yields

$$w = C_1 \int_0^z \xi^{2\theta}(\xi^2 - 1)^{-\theta}\frac{d\xi}{\xi} + C_2. \tag{1.12.76}$$

If $\theta = m/n$ is a rational number, the integral (1.12.76) is rationalized by the substitution $t = (1 - 1/\xi^2)^{1/n}$:

$$w = C \int_0^t \frac{t^{n-m-1}dt}{1 - t^n} + C_2. \tag{1.12.77}$$

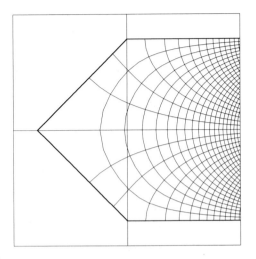

FIGURE 1.105

Example 1.45

When $\theta = 1/4$ ($n = 4$, $m = 1$), $C_2 = 0$ the integral (1.12.77) yields

$$w = C \left(\ln \frac{t+1}{t-1} - 2 \arctan t \right). \tag{1.12.78}$$

The mapping of the Cartesian net of the half-plane $\text{Im } z > 0$ realized by the function (1.12.78) with $t = (1 - 1/z^2)^{1/4}$ is presented in Fig. 1.105.

D. Symmetrical angular domain with the integer angle -2. For the symmetrical quadrangle $(-2, 1 + \theta, 2 - 2\theta, 1 + \theta)$ assume

$$a_1 = \infty, \quad a_2 = -1, \quad a_3 = 0, \quad a_4 = 1;$$
$$\alpha = -2, \quad \beta = 1 + \theta, \quad \gamma = 2 - 2\theta, \quad \delta = 1 + \theta.$$

The Schwarz-Christoffel integral yields

$$w = C_1 \int_{z_0}^{z} (\xi^2 - 1)^\theta \xi^{1-2\theta} d\xi + C_2. \tag{1.12.79}$$

If $\theta = m/n$ is a rational number the integral (1.12.79) is rationalized by the substitution $t = (1 - 1/\xi^2)^{1/n}$:

$$w = C_1 \frac{n}{2} \int_0^t \frac{t^{n+m-1}}{(1-t^n)^2} dt + C_2. \tag{1.12.80}$$

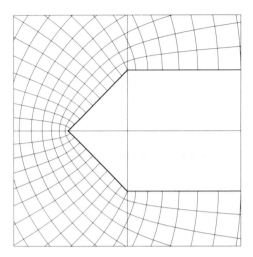

FIGURE 1.106

Example 1.46
When $\theta = 1/4$ ($n = 4$, $m = 1$), the integral (1.12.80) is equal to

$$w = \frac{C_1}{8} \left(\frac{4t}{1 - t^4} - \ln \frac{t+1}{t-1} - 2\arctan t \right) + C_2 . \qquad (1.12.81)$$

The mapping of the Cartesian net of the half-plane Im $z > 0$ carried out by the function (1.12.81) with $t = (1 - 1/z^2)^{1/4}$, $C_1 = 8$, $C_2 = 0$ is presented in Fig. 1.106.

E. Symmetrical quadrangular strip with the integer angle 0. The simplest way to construct the conformal mapping of the symmetrical quadrangle $(0, 1+\theta, -2\theta, 1+\theta)$ onto the strip is to use the symmetry principle. In Subsection 1.12.3 the function $w = f(z_1) = f(e^z)$ was constructed which maps the strip $0 < \text{Im } z < \pi$ onto the triangular strip being a half of the considered quadrangle. At that the real axis Im $z = 0$ is mapped into the axis of symmetry of the quadrangle. Due to the symmetry principle the analytic continuation of the function $w = f(e^z)$ into the strip $-\pi < \text{Im } z < \pi$ conformally maps this strip onto the considered quadrangle.

Example 1.47
The function

$$w(t) = 4t + \ln \frac{t-1}{t+1} - 2\arctan t, \qquad (1.12.82)$$

where $t = (e^z + 1)^{1/4}$, maps the strip $0 < \text{Im } z < \pi$ onto the triangle $(0, -1/4, 5/4)$, the real axis z being mapped into the real axis w. The func-

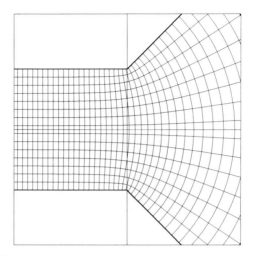

FIGURE 1.107

tion (1.12.82) has the analytic continuation onto the strip $-\pi < \operatorname{Im} z < \pi$ and maps this strip onto the symmetrical quadrangle $(0, 5/4, -1/2, 5/4)$. The mapping of the Cartesian net in the strip $-\pi < \operatorname{Im} z < \pi$ realized by this function is presented in Fig. 1.107.

Exercises

1. When $\theta = 1/4$ ($m = 1$, $n = 4$) the integral (1.12.58) is equal to

$$w = A \left(\arctan t + \operatorname{arth} t - \frac{2t}{p} \right), \qquad (1.12.83)$$

where $A = -2pC_1$. Create the mapping of the strip $0 < \operatorname{Im} z < \pi$ realized by the function (1.12.83) assuming $z_1 = e^z$, $t = (z_1 + 1)^{1/4}$, $A = 1$ for various values of p: $p = 2$, $p = 0.5$, $p = -2$.

2. When $\theta = 1/4$ ($m = 1$, $n = 4$), $p = q^4 > 0$, $0 < q < 1$, $C_1 = 1/2$ the integral (1.12.63) is equal to

$$w = \arctan t + \operatorname{arth} t - \frac{1}{q} \arctan qt - \frac{1}{q} \operatorname{arth} qt . \qquad (1.12.84)$$

Create the mapping of the Cartesian net in the strip $0 < \operatorname{Im} z < \pi$ realized by the function (1.12.84 assuming $z_1 = e^z$, $t = [(z_1 - 1)/(z_1 - q^4)]^{1/4}$, $q = 0.7$.

3. When $\theta = 1/4$ ($m = 1$, $n = 4$), $p = q^4 > 0$, $0 < q < 1$, $C_1 = 2$, $C_2 = 0$ the integral (1.12.68) is expressed as

$$w = 2t \frac{1 - q^4}{1 - t^4} - (1 + 3q^4)(\arctan t + \operatorname{arth} t) + 4q^3(\arctan qt + \operatorname{arth} qt) . \qquad (1.12.85)$$

194 Handbook of Conformal Mapping

Create the mapping of the Cartesian net in the strip $0 < \operatorname{Im} z < \pi$ carried out by the function (1.12.85) assuming $z_1 = e^z$, $t = [(z_1 - 1)/(z_1 - q^4)]^{1/4}$, $q = 0.55$.

4. When $\theta = 3/4$ ($m = 3$, $n = 4$) the integral (1.12.77) is equal to

$$w = \frac{C}{4} \left(2 \arctan t + \ln \frac{t+1}{t-1} \right) + C_2 . \tag{1.12.86}$$

Create the mapping of the Cartesian net in the upper half-plane $\operatorname{Im} z > 0$ realized by the function (1.12.86), where $t = (1 - 1/z^2)^{1/4}$, $C = 4$, $C_2 = -\pi$.

5. When $\theta = 3/4$ ($m = 3$, $n = 4$) the integral (1.12.80) is equal to

$$w = C \left(\frac{4t^3}{1 - t^4} - 3 \ln \frac{t+1}{t-1} + 6 \arctan t \right) + C_2 . \tag{1.12.87}$$

Create the mapping of the Cartesian net in the upper half-plane $\operatorname{Im} z > 0$ carried out by the function (1.12.87) assuming $t = (1 - 1/z^2)^{1/4}$, $C = 1$, $C_2 = -3\pi$.

1.12.5 Mappings of pentagons

Following the terminology of the previous Subsection we will use shortly the term "pentagon $(\alpha, \beta, \gamma, \delta, \epsilon)$" instead of "pentagonal domain with the angles $(\pi\alpha, \pi\beta, \pi\gamma, \pi\delta, \pi\epsilon)$".

Concerning the mapping of a half-plane onto a pentagon with the help of elementary functions considerations similar to those concerning the quadrangles are valid:

1. The Schwarz-Christoffel integral is expressed as elementary functions for pentagons with five integer angles. This is also true for pentagons with three integer and two rational angles. All such pentagons can be enumerated and classified.

2. The Schwarz-Christoffel integral sometimes is expressed as elementary functions for pentagons with two or one integer angle, but these cases are exceptional. It seems impossible to enumerate all these cases.

3. The Schwarz-Christoffel integral is not expressed as elementary functions for polygonal domains without integer angles (in particular, for finite pentagons).

4. The Schwarz-Christoffel integral is expressed as elementary functions for exteriors of finite pentagons only in the case when the pentagon is star-shaped. In this case we have a star-shaped pentagon $(2, 2, \theta, 2, 1 - \theta)$, where $0 < \theta < 1$. In accordance with the formulae of Section 1.12.1 the function mapping the exterior of the circle $|z| > 1$ onto the exterior of the star-shaped cut has the form

$$w = C (z - 1) \left(1 - \frac{a_1}{z} \right)^{\theta} \left(1 - \frac{a_2}{z} \right)^{1-\theta} , \tag{1.12.88}$$

where a_1, a_2 are the points of the unit circle being the preimages of the vertices of angles θ and $1 - \theta$ respectively. The points a_1 and a_2

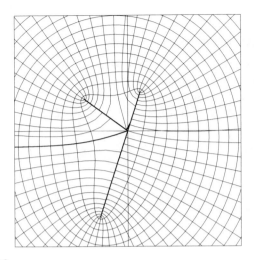

FIGURE 1.108

can be arbitrarily chosen and it is necessary to satisfy the conditions $a_1 \neq 1$, $a_2 \neq 1$, $a_1 \neq a_2$. Varying the values a_1 and a_2 it is possible to obtain any correlation of side lengths of a star-shaped pentagon.

Example 1.48

The mapping of the polar net in the exterior of the circle $|z| > 1$ realized by the function (1.12.88) with $\theta = 0.4$, $a_1 = e^{ip}$, $a_2 = e^{iq}$, $p = \pi/2$, $q = \pi$, $C = 1$ is presented in Fig. 1.108.

Further the classification of all the possible pentagonal domains with five and three integer angles follows.

Pentagons with five integer angles

The only set of five integer angles exists the pentagonal domain can be formed of: two angles 2π, two angles 0 and one angle $-\pi$. Two kinds of pentagons exist for these angles: $(2, 0, 2, 0, -1)$ and $(2, 0, 0, 2, -1)$. Both of these kinds of pentagons are half-planes with two cuts along rays parallel to the edge of the half-plane. Such domains were considered in Subsection 1.12.2 (Examples 1.26 and 1.27).

Pentagons with three integer angles

Let the pentagon have three integer and two non-integer angles $\pi\theta$ and $\pi(k - \theta)$, where k is an integer number. Let the preimage of vertices of non-integer angles be the points of the real axis $z = p$ and $z = q$. The Schwarz-Christoffel integral for the pentagon is

$$w = C_1 \int_{z_0}^{z} R(\xi) \left(\frac{\xi - p}{\xi - q} \right)^{\theta} d\xi + C_2, \qquad (1.12.89)$$

where $R(\xi)$ is a rational function. If $\theta = m/n$ is a rational number the integral (1.12.89) is rationalized by the variable change

$$t = \left(\frac{\xi - p}{\xi - q} \right)^{1/n}, \quad \xi = \frac{p - qt^n}{1 - t^n}.$$

Thus for pentagons with three integer and two rational angles the Schwarz-Christoffel integral is expressed as elementary functions.

All the possible kinds of pentagons with three integer angles can be obtained by considering possible sets of three integer angles, completing them by two angles the sum of all of the angles to be equal to 3π, and considering all the non-trivial permutations of the five obtained angles.

It is possible to arrange 20 various triples from the numbers 2, 0, -1, -2. Ten triples of them can not compose angles of a pentagonal domain. Only the following combinations of integer angles are possible:

$$(2,2,2), \quad (2,2,0), \quad (2,2,-1), \quad (2,2,-2), \quad (2,0,0),$$
$$(2,0,-1), \quad (2,0,-2), \quad (2,-1,-1), \quad (0,0,0), \quad (0,0,-1).$$

The first combination is possible only when the domain is the exterior of a finite star-shaped cut with three rays. This case was considered above. The latest among the enumerated triples of integer angles is possible only when all the angles of a pentagon are integers. These cases also were considered above.

Therefore only 8 combination of the three integer angles are to be investigated. For all the pentagons with 3 integer and 2 rational angles the Schwarz-Christoffel integral is expressed as elementary functions. Everywhere below θ is a rational number lying in the interval $0 < \theta < 1$. We restrict ourselves by enumeration of pentagons without deriving mapping functions.

1. Integer angles $(2,2,0)$. With angles $(2,2,0)$ it is possible to form: the strips $(2,0,2,\theta,-1-\theta)$ and $(2,0,\theta,2,-1-\theta)$ (Fig. 1.109) and the strips with three branches: $(2,0,2,-1+\theta,-\theta)$ and $(2,-1+\theta,2,-\theta,0)$ (Fig. 1.110).

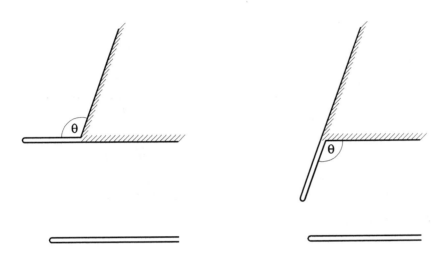

FIGURE 1.109

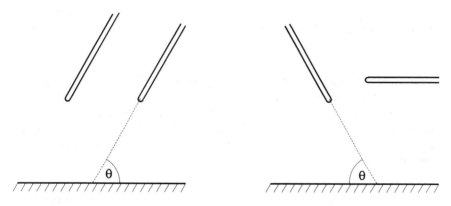

FIGURE 1.110

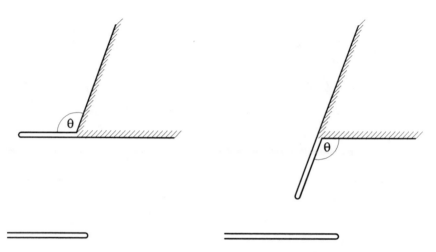

FIGURE 1.111

2. Integer angles $(2, 2, -1)$. Polygonal strips $(2, -1, 2, \theta, -\theta)$ and $(2, -1, \theta, 2, -\theta)$ can be formed with these angles (Fig. 1.111).

3. Integer angles $(2, 2, -2)$. Angular domains $(2, 2, \theta, -2, 1 - \theta)$, $(2, 1 - \theta, \theta, 2, -2)$ and $(2, \theta, 2, -2, 1 - \theta)$ are possible with these angles (Fig. 1.112).

4. Integer angles $(2, 0, 0)$. Following strips are possible with these angles: the strips $(2, \theta, 0, 0, 1 - \theta)$ and $(2, 1 - \theta, \theta, 0, 0)$ (Fig. 1.113), the strip $(2, \theta, 0, 1 - \theta, 0)$ (Fig. 1.114, left), the strip $(2, 0, 1 - \theta, \theta, 0)$ (Fig. 1.114, right), the strips with three branches $(2, -\theta, 1 + \theta, 0, 0)$, $(2, 0, -\theta, 1 + \theta, 0)$ and $(2, 0, 1 + \theta, 0, -\theta)$ (Fig. 1.115).

5. Integer angles $(2, 0, -1)$. It is possible to form the following pentagonal strips with these angles: $(2, \theta, 2 - \theta, 0, -1)$ and $(2, 2 - \theta, 0, -1, \theta)$ (Fig. 1.116, left), $(2, \theta, 0, -1, 2 - \theta)$ and $(2, 0, -1, 2 - \theta, \theta)$ (Fig. 1.116, right), $(2, 0, 2 - \theta, -1, \theta)$ and $(2, \theta, 0, 2 - \theta, -1)$ (Fig. 1.117, left), $(2, -1, \theta, 2 - \theta, 0)$ (Fig. 1.117, right). The dotted line designates the two possible locations of the cut in Fig. 1.116–1.117.

6. Integer angles $(2, 0, -2)$. The only kind of pentagon that exists with such angles is the strip $(2, 0, 2 - \theta, 1 + \theta, -2)$ (Fig. 1.118, left).

7. Integer angles $(2, -1, -1)$. The only kind of pentagon that exists with such angles is the strip $(2, -1, 2 - \theta, 1 + \theta, -1)$ (Fig. 1.118, right).

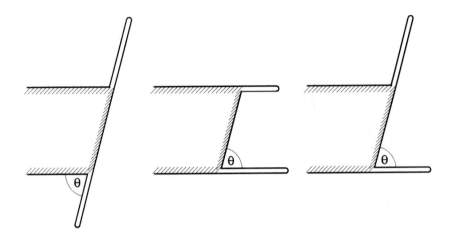

FIGURE 1.112

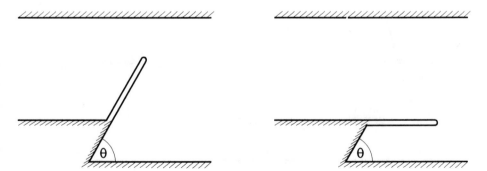

FIGURE 1.113

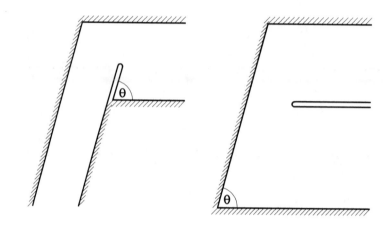

FIGURE 1.114

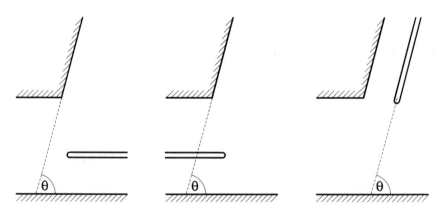

FIGURE 1.115

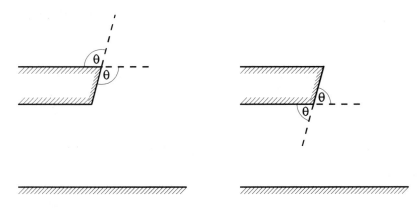

FIGURE 1.116

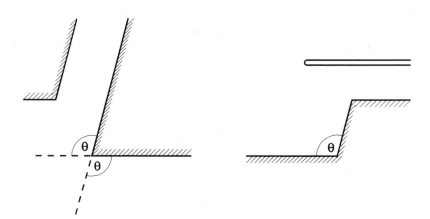

FIGURE 1.117

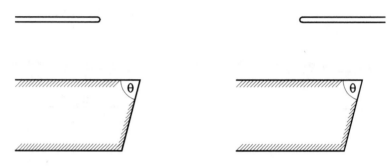

FIGURE 1.118

8. Integer angles $(0,0,0)$. Two kinds of strips with three branches exist for these angles: $(0,0,2-\theta,0,1+\theta)$ and $(0,0,0,2-\theta,1+\theta)$ (Fig. 1.119).

1.12.6   Mappings of regular and symmetrical polygons

1. To construct the conformal mapping of the circle $|z| < 1$ onto the regular n-gon it is necessary to use the formulae $(1.12.1)$–$(1.12.3)$, where a_k are the points of the unit circumference. In the case of the regular polygon the points a_k are also disposed at the vertices of the regular n-gon. It is possible to assume $a_k = \exp(2\pi i k/n)$ $(k = 1, 2, \ldots, n)$ and the formula $(1.12.3)$ yields

$$h(\xi) = \frac{1}{\prod_{k=1}^{n}(\xi - a_k)^{2/n}} \, . \qquad (1.12.90)$$

The numbers a_k are the roots of the equation $\xi^n - 1 = 0$. In accordance with the known algebra theorem the polynomial $\xi^n - 1$ is the product of linear multipliers:

$$\xi^n - 1 = \prod_{k=1}^{n}(\xi - a_k) \, .$$

Thus the function $(1.12.90)$ takes the form

$$h(\xi) = \frac{1}{(\xi^n - 1)^{2/n}} \, .$$

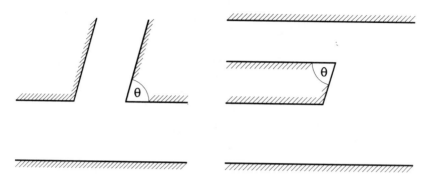

FIGURE 1.119

In order to compute the principal branch of this function continuous inside
the circle $|\xi| < 1$ we should present it as

$$h(\xi) = \frac{A}{(1 - \xi^n)^{2/n}},$$

where $A = \exp 2\pi i/n$. Finally assuming in the formula (1.12.1) $C_1 A = C$,
$C_2 = 0$ the mapping function is represented as

$$w = C \int_0^z \frac{d\xi}{(1 - \xi^n)^{2/n}}.$$

(1.12.91)

The function (1.12.91) is not the elementary. For its calculation it is possible to expand the integrand function into a Taylor series and to integrate
it termwise. Assuming $C = 1$ the following series is obtained:

$$w = z + \frac{2}{n}\frac{z^{n+1}}{n+1} + \frac{2}{n}\left(\frac{2}{n}+1\right)\frac{1}{2!}\frac{z^{2n+1}}{2n+1}$$

$$+ \frac{2}{n}\left(\frac{2}{n}+1\right)\left(\frac{2}{n}+2\right)\frac{1}{3!}\frac{z^{3n+1}}{3n+1} + \cdots.$$

(1.12.92)

In Fig. 1.120 the mapping of the polar net inside the circle $|z| < 1$ realized
by the function (1.12.92) with $n = 5$ and 8 terms kept is represented.

2. To construct the conformal mapping of the exterior of the circle
$|z| > 1$ onto the exterior of the regular n-gon it is necessary to use the
formula (1.12.5). In this case the points a_k are also disposed at the unit
circle in the vertices of the regular n-gon. Assume $a_k = \exp(2\pi i k/n)$ and

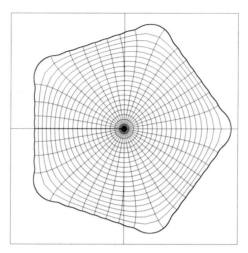

FIGURE 1.120

the function (1.12.5) yields

$$w = C_1 \int_{z_0}^{z} g(\xi)\, d\xi + C_2 , \qquad (1.12.93)$$

where $g(\xi) = 1/\xi^2 \prod_{k=1}^{n}(\xi - a_k)^{2/n} = 1/\xi^2\,(\xi^n - 1)^{2/n} = (1 - 1/\xi^n)^{2/n}$. To compute the non-elementary function (1.12.93) it is possible to expand function $g(\xi)$ into the Laurent series when $|\xi| > 1$ and to integrate term by term. Assuming $C_1 = 1$, $C_2 = 0$ we obtain

$$w = z + \frac{2}{n}\frac{z^{1-n}}{1-n} - \frac{2}{n}\left(\frac{2}{n} - 1\right)\frac{1}{2!}\frac{z^{1-2n}}{2n-1} + \cdots . \qquad (1.12.94)$$

The mapping of the polar net in the exterior of the circle $|z| > 1$ realized by the function (1.12.94) with $n = 3$ and three terms kept is presented in Fig. 1.121

3. Let the domain G be the half-plane $\operatorname{Im} w > 0$ with $(n-1)$ symmetrical cuts along segments $[0, h\exp(\pi i k/n)]$ $(k = 1, 2, \ldots, n-1)$ with the ends lying at the vertices of the regular $2n$-gon. To map the domain G into the half-plane $\operatorname{Im} \xi > 0$ it is necessary to use the function which maps the exterior of the circle $|z| > 1$ onto the exterior of star-shaped cut consisting of $2n$ equal segments (see Subsection 1.10.8):

$$w = \left[\frac{1}{2}\left(z^n + \frac{1}{z^n}\right)\right]^{1/n} . \qquad (1.12.95)$$

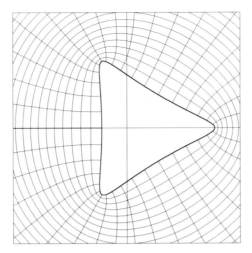

FIGURE 1.121

The function (1.12.95) takes real values with real values of z if $|z| > 1$. Thus it maps the half-plane Im $z > 0$ with the half-circle $|z| < 1$ deleted onto the considered domain G. The half-plane Im $\xi > 0$ is mapped onto the half-plane Im $z > 0$ with the half-circle $|z| < 1$ deleted with the help of the inverse Zhukovskii function:

$$z = \xi + \sqrt{\xi^2 - 1}.$$

Therefore the mapping of the half-plane Im $\xi > 0$ onto the domain G is realized by the composition of functions

$$z = \xi + \sqrt{\xi^2 - 1},$$

$$w = \left[\frac{1}{2} \left(z^n + \frac{1}{z^n} \right) \right]^{1/n}. \tag{1.12.96}$$

The function (1.12.96) can be presented as

$$w = (T_n(\xi))^{1/n}, \tag{1.12.97}$$

where

$$T_n(\xi) = \frac{1}{2} [(\xi + \sqrt{\xi^2 - 1})^n + (\xi - \sqrt{\xi^2 - 1})^n]. \tag{1.12.98}$$

The function T_n is a polynomial called the *Chebyshev polynomial*. Another representation is

$$T_n(\xi) = \cos(n \arccos \xi) = \cosh(n \operatorname{arch} \xi).$$

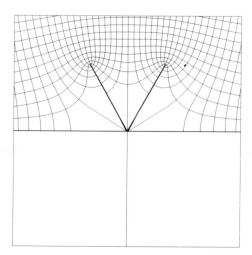

FIGURE 1.122

To compute the principal branch of the function (1.12.97) being continuous in the half-plane Im $\xi > 0$ we should represent it as

$$w = \xi \left(\frac{T_n(\xi)}{\xi^n} \right)^{1/n} = \xi \left[\frac{\cos(n \arccos \xi)}{\xi^n} \right]^{1/n}. \tag{1.12.99}$$

The mapping of the Cartesian net in the half-plane Im $\xi > 0$ realized by the function (1.12.99) with $n = 3$ is presented in Fig. 1.122

   4.   The mapping of the circle $|z| < 1$ onto the strip with n symmetrical straight branches is carried out by the composition of functions

$$t = z \left(\frac{2}{1 + z^n} \right)^{2/n}, \tag{1.12.100}$$

$$w = C_1 \int \frac{dt}{t^n - 1} + C_2. \tag{1.12.101}$$

The integral (1.12.101) is calculated by the expansion of the integrand function into the partial fractions:

$$\frac{1}{t^n - 1} = \frac{1}{n} \sum_{k=1}^{n} \frac{a_k}{t - a_k},$$

where $a_k = \exp(2\pi i k/n)$. The integral (1.12.101) is

$$w = \overline{C_1} \sum_{k=1}^{n} a_k \ln(t - a_k) + C_2.$$

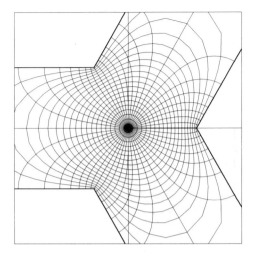

FIGURE 1.123

To compute the principal branch of this multi-valued function, which is continuous inside the circle $|z| < 1$, it should be presented as

$$w = \overline{C_1} \sum_{k=1}^{n} a_k \ln \left(1 - \frac{t}{a_k} \right) + C .$$

The mapping of the unit circle realized by the composition of functions

$$t = z \left(\frac{2}{1 + z^n} \right)^{2/n}$$

$$w = \sum_{k=1}^{n} a_k \ln \left(1 - \frac{t}{a_k} \right)$$

with $n = 3$ is presented in Fig. 1.123.

1.13 Mappings of Rectangular Domains. Elliptic Integrals and Elliptic Functions

1.13.1 Mappings of rectangles. The elliptic integral of the first kind

Consider a rectangle in the complex plane w, which is symmetrical with respect to the imaginary axis (Fig. 1.124, a).

$$-b < \operatorname{Re} w < b, \quad 0 < \operatorname{Im} w < H .$$

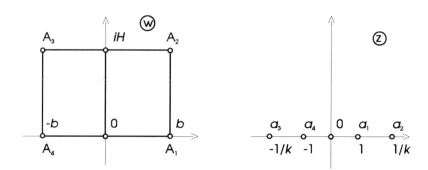

FIGURE 1.124

In accordance with the Riemann's theorem it can be conformally mapped (uniquely) onto the half-plane $\operatorname{Im} z > 0$, the boundary points $w = -b, 0, b$ being mapped into the boundary points $z = -1, 0, 1$ respectively (Fig. 1.124, b). Let the image of the point A_2 ($w_2 = b + iH$) be a point of the real axis $a_2 = 1/k$; due to the principle of symmetry the image of the point A_3 ($w_3 = -b + iH$) is the point $a_3 = -1/k$.

Assume in the formulae (1.12.1)–(1.12.2)

$$\begin{aligned}
a_1 &= 1, & a_2 &= 1/k & a_3 &= -1/k & a_4 &= -1; \\
\alpha &= 1/2, & \beta &= 1/2, & \gamma &= 1/2, & \delta &= 1/2; \\
z_0 &= 0, & C_2 &= w(z_0) = 0.
\end{aligned}$$

The function realizing the inverse mapping of the half-plane $\operatorname{Im} z > 0$ onto the rectangle is

$$w = C_1 \int_0^z \frac{d\xi}{\sqrt{(1 - \xi^2)(1 - k^2\xi^2)}} . \tag{1.13.1}$$

The function

$$F(z, m) = \int_0^z \frac{d\xi}{\sqrt{(1 - \xi^2)(1 - m\xi^2)}} . \tag{1.13.2}$$

is called the *elliptic integral of the first kind*.*

---

*In the general case the *elliptic integral* is $\int R(z, w)\, dz$, where $R(z, w)$ is a rational function of two variables, $w = \sqrt{P(z)}$, $P(z)$ is a polynomial of third or fourth power without multiple roots. Generally speaking elliptic integrals are not expressed as ele-

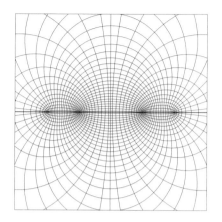 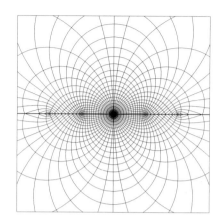

FIGURE 1.125

The number m is called the *parameter* and the number $k = \sqrt{m}$ — the *modulus of the elliptic integral*. Provisionally suppose that $0 < m < 1$.

The function $F(z, m)$ is an infinitely-valued analytic function with the branch points $z = \pm 1$, $z = \pm 1/\sqrt{m}$. It is possible to separate its singly-valued continuous branch if to make cuts along the rays of the real axis $(-\infty, -1)$ and $(1, \infty)$. Only this principal branch will be considered further. This function has real values on the segment $(-1, 1)$.

The isothermic net and the map of the relief of the function $w = F(z, m)$ is presented in Fig. 1.125, a and 1.125, b for $m = 0.25$.

The value

$$F(1, m) = \int_0^1 \frac{dx}{\sqrt{(1 - x^2)(1 - mx^2)}} = \int_0^{\pi/2} \frac{d\varphi}{\sqrt{1 - m\sin^2\varphi}} \qquad (1.13.3)$$

is called the *complete elliptic integral of the first kind* and is designated as $K(m)$ or simply K. Thus the segment $(-1, 1)$ is mapped into the segment $(-K, K)$ by the function (1.13.2).

The function $w = F(z, m)$ is univalent in the upper half-plane $\text{Im } z > 0$ and conformally maps it onto some rectangle with the segment $(-K, K)$

---

mentary functions. In particular the integral (1.13.2) is an elementary function only when $m = 0$ or $m = 1$:

$$F(z, 0) = \arcsin z, \quad F(z, 1) = \text{arth } z.$$

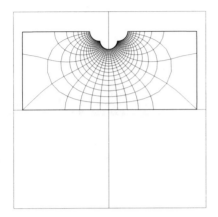 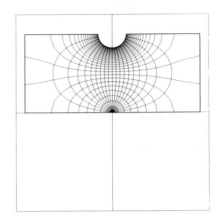

FIGURE 1.126

being its base. Here the point $a_2 = 1/k = 1/\sqrt{m}$ is mapped into the point

$$w = F(1/k, m) = \int_0^1 \frac{dx}{\sqrt{(1-x^2)(1-mx^2)}} + i \int_1^{1/k} \frac{dx}{\sqrt{(x^2-1)(1-mx^2)}} \tag{1.13.4}$$

The first term in the formula (1.13.4) is the complete elliptic integral $K(m)$ and the second term is transformed by the change of the variable

$$mx^2 + m't^2 = 1, \tag{1.13.5}$$

where $m' = 1 - m$ is the so-called *complementary parameter of the elliptic integral*. After the change (1.13.5) we obtain

$$\int_1^{1/k} \frac{dx}{\sqrt{(x^2-1)(1-mx^2)}} = \int_0^1 \frac{dt}{\sqrt{(1-t^2)(1-m't^2)}} .$$

For the brevity the value of the complete elliptic integral of the first kind of the complementary parameter $K(m')$ is designated as K'. Therefore the point $a_2 = 1/k = 1/\sqrt{m}$ is mapped into the point $K + iK'$ by the function $F(z, m)$, and the whole upper half-plane onto the rectangle with the base $2K$ and the height K'.

The mapping of the Cartesian net in the half-plane Im $z > 0$ carried out by the function $w = F(z, m)$ with $m = 0.64$ is presented in Fig. 1.126, a. In Fig. 1.126, b the mapping of the polar net in the upper half-plane realized by the same function is presented. The lower half-plane Im $z < 0$ is mapped by the function $w = F(z, m)$ onto the rectangle symmetrical to that depicted in Fig. 1.126 with respect to the real axis.

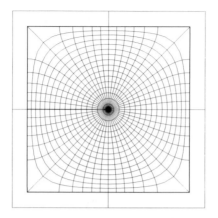

FIGURE 1.127

To construct the mapping of the circle onto the rectangle it is first necessary to map the circle onto the plane with cuts along rays of the real axis
$(-\infty, -1)$ and $(1, \infty)$. It is realized by the function

$$t = \frac{2z}{1 + z^2} \,. \tag{1.13.6}$$

Further the plane t with cuts along the mentioned rays is mapped onto the
rectangle by the function

$$w = F(t, m) \,. \tag{1.13.7}$$

The mapping of the polar net in the circle $|z| < 1$ onto the square car-
ried out by the function (1.13.6)–(1.13.7) with $m = 1/2$ is presented in
Fig. 1.127.

The parameter m is unknown when searching for the function (1.13.1),
which maps the half-plane onto the given rectangle. It can be found from
the equation

$$\frac{K(1 - m)}{K(m)} = \frac{H}{b} \,,$$

then the constant C_1 is determined as $b/K(m)$.

The values of functions $K(m)$, $K'(m)$, $K'(m)/K(m)$ for various values
of the parameter m are given in Table 1.1. With the help of this Table it is possible to define m approximately as the function of the ratio
$K'(m)/K(m)$ by mean of interpolation. For small values of m the values
$K(m)$ and $K'(m)$ can be evaluated by the asymptotical formulae

$$K(m) = \frac{\pi}{2} \left(1 + \frac{m}{4} + O(m^2)\right) \,,$$

TABLE 1.1

m	$K(m)$	$K'(m)$	$K'(m)/K(m)$
0.01	1.575	3.696	2.35
0.05	1.591	2.908	1.83
0.10	1.612	2.578	1.60
0.20	1.660	2.257	1.36
0.30	1.714	2.075	1.21
0.40	1.777	1.950	1.10
0.50	1.854	1.854	1.00
0.60	1.950	1.777	0.91
0.70	2.075	1.714	0.83
0.80	2.257	1.660	0.74
0.90	2.578	1.612	0.63
0.95	2.908	1.591	0.55
0.99	3.696	1.575	0.43

$$K'(m) = \frac{1}{2}\left(1 + \frac{m}{4} + O(m^2)\right)\ln\frac{16}{m} - \frac{m}{4}.$$

When building the mappings with the help of the program CONFORM it is more convenient to adjust the parameter m experimentally in order to obtain the desired shape of the rectangle.

1.13.2 Mapping of infinite rectangular domain. The elliptic integral of the second kind

Let an infinite pentagonal domain of the plane w be the half-plane Im $z >$ H with rectangular projection $-b <$ Re $w < b$, $0 <$ Im $w < H$, which is symmetrical with respect to the imaginary axis (Fig. 1.128, a). Due to the Riemann's theorem this domain can be conformally mapped (uniquely) onto the half-plane Im $z > 0$, so that the boundary points $w = -b, 0, b$ would be mapped into the points $z = -1, 0, 1$ correspondingly. Let the point $w_2 = b + iH$ be mapped into the point of the real axis $a_2 = 1/k$, then the image of the point A_4 ($w_4 = -b + iH$) will be the point $a_4 = -1/k$ ($0 < k < 1$) in accordance with the symmetry principle. Assume in the formulae (1.12.1)–(1.12.2)

$$a_1 = 1, \qquad a_2 = 1/k, \qquad a_3 = \infty, \quad a_4 = -1/k, \quad a_5 = -1;$$
$$\alpha_1 = 1/2, \qquad \alpha_2 = 3/2, \qquad \alpha_3 = -1, \quad \alpha_4 = 3/2 \quad \alpha_5 = 1/2;$$
$$z_0 = 0, \qquad C_2 = w(z_0) = 0.$$

The Schwarz-Christoffel integral yields

$$w = C_1 \int_0^z \sqrt{\frac{1 - k^2\xi^2}{1 - \xi^2}}\, d\xi. \qquad (1.13.8)$$

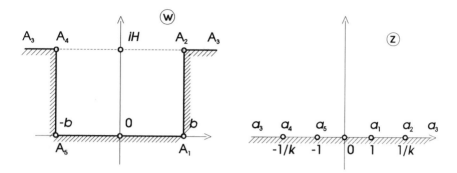

FIGURE 1.128

The function

$$F(z, m) = \int_0^z \sqrt{\frac{1 - m\xi^2}{1 - \xi^2}} \, d\xi. \tag{1.13.9}$$

is called the *elliptic integral of the second kind*, the number $k = \sqrt{m}$ is called its modulus (temporarily we consider the values of parameters m from the interval $0 < m < 1$). The integral (1.13.9) is an infinitely-valued non-elementary function with branch points $z = \pm 1$, $z = \pm 1/k = \pm 1/\sqrt{m}$. Consider only its principal branch, which is continuous and analytic in the plane z with cuts along rays of the real axis $(-\infty, -1)$ and $(1, \infty)$. Like the elliptic integral of the first kind the function $E(z, m)$ is an odd function taking real values on the segment $(-1, 1)$.

The isothermic net of the function $E(z, m)$ with $m = 0.25$ is presented in Fig. 1.129, a. In Fig. 1.129, b the relief map of the same function is presented.

The value

$$E(1, m) = \int_0^1 \sqrt{\frac{1 - mx^2}{1 - x^2}} \, dx = \int_0^{\pi/2} \sqrt{1 - m \sin^2 \varphi} \, d\varphi$$

is called the *complete elliptic integral of the second kind* and is designated as $E(m)$ or simply E (not to be confused with the function $E(z, m)$ of two arguments!). Therefore the function $w = E(z, m)$ maps the segment of the real axis $(-1, 1)$ into the segment $(-E, E)$.

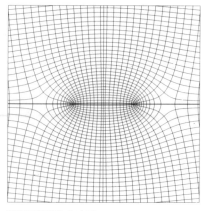 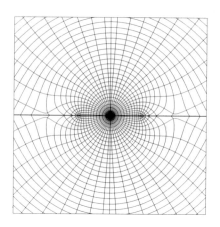

FIGURE 1.129

The function $E(z, m)$ is univalent in all the domain of definition. It conformally maps the upper half-plane Im $z > 0$ onto some polygonal domain, the point $z = 1/k = 1/\sqrt{m}$ is mapped into the point

$$E(1/k, m) = \int_0^1 \sqrt{\frac{1 - mx^2}{1 - x^2}}\, dx + i \int_1^{1/k} \sqrt{\frac{1 - mx^2}{x^2 - 1}}\, dx. \qquad (1.13.10)$$

The first term in the formula (1.13.10) is the complete elliptic integral of the second kind $E(m)$ and the second term is reduced by the change of the variable (1.13.5)

$$t = \sqrt{\frac{1 - mx^2}{m'}}, \qquad x = \sqrt{\frac{1 - m't^2}{m}},$$

where $m' = 1 - m$ is the complementary parameter. After the change of the variable

$$\int_1^{1/k} \sqrt{\frac{1 - mx^2}{x^2 - 1}}\, dx = \int_0^1 \frac{m't^2\, dt}{\sqrt{(1 - t^2)(1 - m't^2)}} = K(m') - E(m').$$

The value $E(m')$ for brevity is designated as E'. Thus the point $z_2 = 1/k = 1/\sqrt{m}$ is mapped into the point $w_2 = E + i(K' - E')$. All the half-plane Im $z > 0$ is mapped by the function $w = E(z, m)$ onto the pentagonal angular domain, which is the half-plane Im $w > K' - E'$ with the rectangular projection $-E < \operatorname{Re} w < E$, $0 < \operatorname{Im} w < K' - E'$.

The mapping of the Cartesian net in the upper half-plane Im $z > 0$ carried out by the function $w = E(z, m)$ with $m = 0.25$ is presented in Fig. 1.130, a. The plane Im $z < 0$ is mapped by the function $w = E(z, m)$ onto the domain symmetrical to one depicted in Fig. 1.130, a with respect

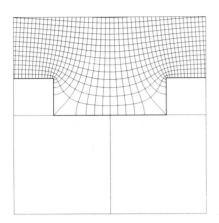 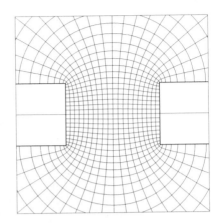

FIGURE 1.130

to the real axis. All the plane z with cuts along rays of the real axis $(-\infty, -1)$ and $(1, \infty)$ is conformally mapped onto the polygonal strip. In order to map the adequate canonical domain — the strip $0 < \mathrm{Im}\, z < \pi$ — onto the considered polygonal domain the intermediate variable should be used

$$t = \cosh z, \quad w = E(t, m). \tag{1.13.11}$$

The mapping of the Cartesian net in the strip $0 < \mathrm{Im}\, z < \pi$ realized by the function (1.13.11) with $m = 0.25$ is presented in Fig. 1.130, b. When building the mapping of the half-plane onto the given infinite rectangular domain (Fig. 1.128, a) the parameter m is unknown. It is determined from the equation

$$\frac{K'(m) - E'(m)}{E(m)} = \frac{H}{b}. \tag{1.13.12}$$

The values of the function $[K'(m) - E'(m)]/E(m)$ for some values of m are presented in Table 1.2. With the help of this Table it is possible to define the root of the equation (1.13.12) by mean of interpolation. When creating the mapping with the help of the program CONFORM the parameter m is defined experimentally.

When the parameter m is greater than 1 the elliptic integrals (1.13.2) and (1.13.9) are reduced by the change of the integration variable $\tau = \sqrt{m}\xi$ to the elliptic integrals with the parameter less than 1. Designating $p = 1/m$ (where $0 < p < 1$) we obtain

$$F(z, 1/p) = \sqrt{p}\, F(z/\sqrt{p}, p),$$

TABLE 1.2

m	$E(m)$	$E'(m)$	$(K'-E')/E$
0.01	1.567	1.016	1.71
0.05	1.551	1.060	1.19
0.10	1.531	1.105	0.96
0.20	1.489	1.178	0.72
0.30	1.445	1.242	0.58
0.40	1.399	1.298	0.47
0.50	1.351	1.351	0.37
0.60	1.298	1.399	0.29
0.70	1.242	1.445	0.22
0.80	1.178	1.489	0.15
0.90	1.105	1.531	0.07
0.95	1.060	1.551	0.04
0.99	1.016	1.567	0.01

$$E(z,1/p) = \frac{1}{\sqrt{p}}\left[E\left(\frac{z}{\sqrt{p}},p\right) - (1-p)F\left(\frac{z}{\sqrt{p}},p\right)\right].$$

The function $w = F(z,m)$ with $m > 1$ conformally maps the half-plane $\operatorname{Im} z > 0$ onto the rectangle with the base $2\sqrt{p}\,K(p)$ and the height $\sqrt{p}\,K'(p)$.

The function $w = E(z,m)$ with $m > 1$ conformally maps the half-plane $\operatorname{Im} z > 0$ onto the half-plane

$$\operatorname{Im} w > -\frac{E'(p) - p\,K'(p)}{\sqrt{p}}$$

with the deleted rectangle

$$|x| < \frac{E(p) - p'\,K(p)}{\sqrt{p}}, \qquad -\frac{E'(p) - p\,K'(p)}{\sqrt{p}} < y < 0, \qquad (1.13.13)$$

where $p = 1/m$, $p' = 1 - p = 1 - 1/m$.

The mapping of the Cartesian net in the half-plane $\operatorname{Im} z > 0$ carried out by the function $w = E(z,m)$ with $m = 2$ is presented in Fig. 1.131. In this case the domain is the half-plane with the rectangle deleted, which is a half of a square.

1.13.3 Mappings of quadrangles with four right angles

Infinite rectangles with four right angles are divided into three types

1. Angular domains with angles $\pi/2$, $\pi/2$, $3\pi/2$, $-\pi/2$,
2. Rectangular strips with angles $3\pi/2$, $3\pi/2$, $-\pi/2$, $-\pi/2$,

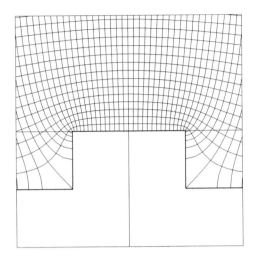

FIGURE 1.131

3. Exteriors of rectangles.

All these rectangles can be represented as conformal mappings of canonical domains with the help of the elliptic integrals.

1. Angular domains with four right angles are rectangles $(1/2, 1/2, 3/2, -1/2)$ and $(1/2, 3/2, 1/2, -1/2)$. These domains are obtained under the mapping of the first quadrant of the plane t by the function $w = E(t, m)$ with $0 < m < 1$ or with $m > 1$. In order to map the half-plane Im $z > 0$ onto the considered domain it is necessary to use t as an intermediate variable: $t = \sqrt{z}$.

Mappings of the Cartesian net of the half-plane Im $z > 0$ realized by the function

$$t = \sqrt{z}, \quad w = E(t, m)$$

are presented in Fig. 1.132 a, b with $m = 0.5$ and $m = 1.5$.

2. Rectangular strips with right angles are rectangles $(-1/2, 3/2, 3/2, -1/2)$ and $(-1/2, 3/2, -1/2, 3/2)$. The first of these rectangles can be considered as the half of the hexagon shown in Fig. 1.130, b. To map the canonical domain — the strip $0 < \text{Im } z < \pi$ — onto the considered rectangular strip it is necessary to use the intermediate variable $t = \cosh(z/2)$.

The mapping of the Cartesian net in the strip $0 < \text{Im } z < \pi$ realized by the function

$$t = \cosh(z/2), \quad w = E(z, m)$$

with $m = 0.4$ is presented in Catalog 4, Domain 36.

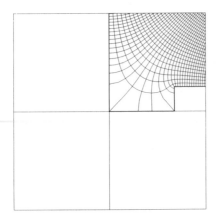 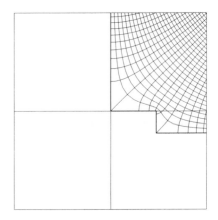

FIGURE 1.132

The mapping of the same strip onto the rectangle $(-1/2, 3/2, -1/2, 3/2)$ realized by the sequence of transformations

$$u = e^z ,$$

$$t = \sqrt{u + 1} ,$$

$$s = t\sqrt{m - m'/u} ,$$

$$w = s + F(t, m) - 2E(t, m)$$

is presented for $m = 0.5$ in Catalog 4, Domain 39.

3. The mapping of the exterior of a circle onto the exterior of a rectangle can be obtained on the basis of the symmetry principle using the mapping of the half-plane with the rectangle (1.13.13) deleted. The function

$$w(t) = E\left(t, \frac{1}{p}\right) + i\,\frac{E'(p) - p\,K'(p)}{\sqrt{p}} , \qquad (0 < p < 1) \qquad (1.13.14)$$

maps the half-plane Im $t > 0$ onto the half-plane Im $w > 0$, it is real-valued on the rays of the real axis $(-\infty, -1)$ and $(1, \infty)$. The function (1.13.14) allows the analytic continuation across these rays into the lower half-plane Im $t < 0$, which is a singly-valued analytical function in the plane t with the cut along segment $(-1, 1)$. The conformal mapping of the exterior of the circle $|z| > 1$ onto the exterior of the segment $(-1, 1)$ is carried out by the Zhukovskii function

$$t = \frac{1}{2}\left(z + \frac{1}{z}\right) .$$

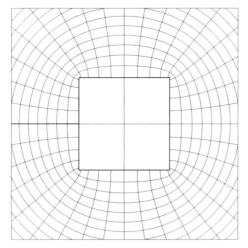

FIGURE 1.133

Finally the conformal mapping of the exterior of the circle $|z| > 1$ onto
the exterior of the rectangle is carried out by the composition of functions

$$t = \frac{1}{2}\left(z + \frac{1}{z}\right),$$

$$A = \frac{E'(p) - p\,K'(p)}{\sqrt{p}}, \tag{1.13.15}$$

$$w = E(t, 1/p) + i\,A\,\mathrm{sign}\,(\mathrm{Im}\,t),$$

where $0 < p < 1$.

The mapping of the polar net in the exterior of the circle $|z| > 1$ realized
by the function (1.13.15) with $p = 1/2$ is presented in Fig. 1.133. In this
case the domain represents the exterior of the square.

1.13.4   Mappings realized by linear combinations of elliptic integrals

Consider mappings carried out by the function

$$w = E(z, m) - A\,F(z, m) \tag{1.13.16}$$

with the real values of A. The parameter m without loss of generality can
be assumed to lie in the interval $0 < m < 1$.

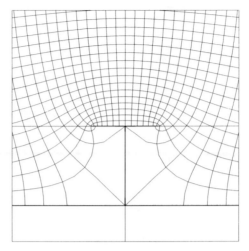

FIGURE 1.134

The function (1.13.16) is an odd one having real values at the interval $-1 < x < 1$. The values of the function at the ends of the interval $(0, 1)$ are

$$w(0) = 0, \quad w(1) = E - AK.$$

For the mapping (1.13.16) to be univalent it is necessary that $w(1) \geq 0$, i.e., $A \leq E(m)/K(m)$.

If $A = E/K$ the function (1.13.16) conformally maps the half-plane $\text{Im } z > 0$ onto the half-plane $\text{Im } w > (1 - E/K)K' - E'$ with a T-shaped cut.

The mapping of the Cartesian net of the half-plane $\text{Im } z > 0$ realized by the function

$$w = E(z, m) - \frac{E(m)}{K(m)} F(z, m) \tag{1.13.17}$$

with $m = 0.81$ is presented in Fig. 1.134.

The change of the shape of the domain obtained by the mapping of the upper half-plane with the help of the function (1.13.16) (under the change of the parameter A) is shown in Fig. 1.135, a–h. If $A > E/K$ the function (1.13.16) is not univalent in the upper half-plane.

On the basis of the symmetry principle using the mappings presented in Fig. 1.135 it is possible to construct mappings of some polygonal strips and exteriors of finite polygons.

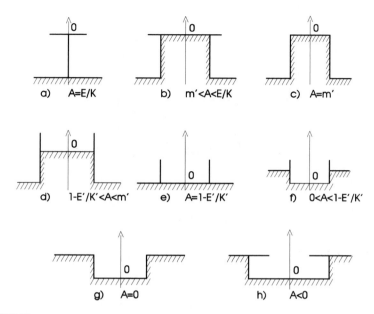

FIGURE 1.135

Example 1.49
The mapping of the Cartesian net in the strip $0 < \text{Im } z < \pi$ carried out by the function

$$t = \cosh z \,,$$

$$w = E(t, m) - A\,F(t, m)$$

is shown in Fig. 1.136. It was built with $m = 0.1$, $A = 0.45$. The mappings realized by this function with other values of the parameters are presented in Catalog 4 (Domains 37 and 38).

Example 1.50
The mapping of the exterior of the circle $|z| > 1$ onto the exterior of the H-shaped cut is realized by the composition of functions

$$t = \frac{1}{2\sqrt{m}} \left(z + \frac{1}{z} \right) \,,$$

$$A = 1 - \frac{E(1 - m)}{K(1 - m)} \,,$$

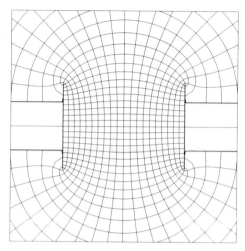

FIGURE 1.136

$$w = E(t,m) - A\,F(t,m)\,.$$

The mapping of the polar net realized by this function with $m = 0.1$ is presented in Catalog 2 (Domain 26).

1.13.5 Elliptic Jacobian functions

The function inverse to the elliptic integral of the first kind $F(z,m)$ is called the *elliptic sine* and is designated as $\operatorname{sn}(z,m)$. Two other functions are closely connected with the elliptic sine, namely the *elliptic cosine*

$$\operatorname{cn}(z,m) = \sqrt{1 - \operatorname{sn}^2(z,m)}\,,\quad \operatorname{cn}(0,m) = 1$$

and *delta amplitude*

$$\operatorname{dn}(z,m) = \sqrt{1 - m\,\operatorname{sn}^2(z,m)}\,,\quad \operatorname{dn}(0,m) = 1\,.$$

The functions sn, cn and dn are called the *basic elliptic Jacobian functions*.

Elliptic Jacobian functions are elementary functions of the variable z only when $m = 0$ or $m = 1$:

$$\begin{aligned}
\operatorname{sn}(z,0) &= \sin z\,, & \operatorname{cn}(z,0) &= \cos z\,, & \operatorname{dn}(z,0) &= 1\,; \\
\operatorname{sn}(z,1) &= \tanh z\,, & \operatorname{cn}(z,1) &= \operatorname{sech} z\,, & \operatorname{dn}(z,1) &= \operatorname{sech} z\,.
\end{aligned}$$

We consider only real values of the parameter m: $0 < m < 1$. For these values of m elliptic Jacobian functions are non-elementary meromorphic doubly periodic functions of a complex variable z.

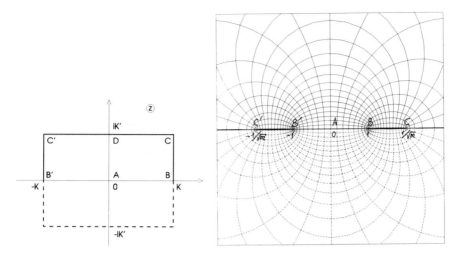

FIGURE 1.137

The function $w = \operatorname{sn}(z, m)$ realizes the conformal mapping inverse to that carried out by the function $z = F(w, m)$, namely it maps the rectangle

$$-K(m) < \operatorname{Re} z < K(m), \quad 0 < \operatorname{Im} z < K'(m) \qquad (1.13.18)$$

onto the upper half-plane $\operatorname{Im} w > 0$ and

$$\begin{aligned}
\operatorname{sn}(0, m) &= 0, & \operatorname{sn}(\pm K, m) &= \pm 1, \\
\operatorname{sn}(\pm K + iK', m) &= \pm 1/\sqrt{m}, & \operatorname{sn}(iK', m) &= \infty.
\end{aligned}$$

The rectangle (1.13.18) is called the *basic rectangle* for elliptic Jacobian functions with the given parameter m. The functions $\operatorname{cn}(z, m)$ and $\operatorname{dn}(z, m)$ are also univalent in the basic rectangle.

The analytic continuation of the function $w = \operatorname{sn}(z, m)$ is the single-valued odd doubly periodic analytic function with the periods $4K(m)$ and $2iK'(m)$. The basic rectangle represents the quarter of the period of this doubly periodic function.

The mapping of the Cartesian net in the basic rectangle carried out by the function $w = \operatorname{sn}(z, m)$ with $m = 0.25$ is shown in Fig. 1.137, b by solid lines.

The function $w = \operatorname{sn}(z, m)$ is univalent in the doubled basic rectangle

$$-K < \operatorname{Re} z < K, \quad -K' < \operatorname{Im} z < K'$$

and maps it conformally onto the plane w with the cut along rays of the real axis $(-\infty, -1)$ and $(1, \infty)$. The mapping of the Cartesian net in this rectangle is shown in Fig. 1.137, b by solid and dashed lines. It is necessary to mention that the net of lines is the isothermic net of the inverse function $w = F(z, 0.25)$.

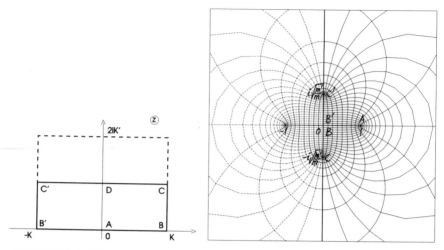

FIGURE 1.138

The isothermic net of the function $w = \text{sn}\,(z, m)$ with $m = 0.25$ in the basic rectangle is shown in Fig. 1.126, a. The relief map of this function is presented in Fig. 1.126, b.

The function $w = \text{sn}\,(z, m)$ has zeroes of the first order at the points $z_{nl} = 2nK + i \cdot 2lK'$ and poles of the first order at the points $z_{nl}^* = 2nK + i(2l+1)K'$ $(n, l \in \mathcal{Z})$. The basic elliptic Jacobian functions $\text{cn}\,(z, m)$ and $\text{dn}\,(z, m)$ have first order poles at the same points z_{nl}^*.

The Jacobian function $w = \text{cn}\,(z, m)$ is the even doubly periodic singly-valued meromorphic function with the periods $4K$ and $2K + i \cdot 2K'$. This function has zeroes of the first order at the points $(2n + 1)K + i \cdot 2lK'$ $(n, l \in \mathcal{Z})$. Its particular values are:

$$\text{cn}\,(0, m) = 1, \qquad\qquad \text{cn}\,(\pm K, m) = 0,$$
$$\text{cn}\,(\pm K + iK', m) = \pm i\sqrt{m'/m}, \quad \text{cn}\,(\pm iK', m) = \infty.$$

The function $w = \text{cn}\,(z, m)$ conformally maps the basic rectangle onto the right half-plane $\text{Re}\,w > 0$ with the cut along the segment $[0, 1]$. The mapping of the Cartesian net in the basic rectangle by the function $w = \text{cn}\,(z, m)$ with $m = 0.6$ is shown in Fig. 1.138, b by solid lines.

The function $w = \text{cn}\,(z, m)$ is univalent in the doubled basic rectangle $-K < x < K$, $0 < y < 2K'$ and conformally maps it onto the plane w with cross-shaped cut along the segments $[-1, 1]$ and $[-i\sqrt{m'/m}, i\sqrt{m'/m}]$. The mapping of the Cartesian net in the doubled rectangle is shown in Fig. 1.138, b by solid and dashed lines.

The Jacobian delta amplitude function $w = \text{dn}\,(z, m)$ is also the even doubly periodic singly-valued meromorphic function with the periods $2K$ and $4iK'$. This function has zeroes of the first order at the points $(2n +$

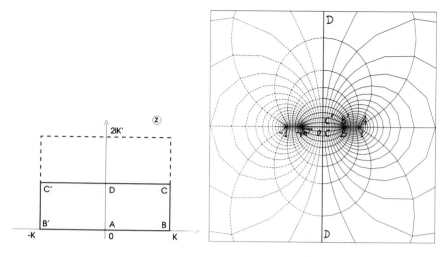

FIGURE 1.139

1)$K + i(2l + 1)K'$ $(n, l \in \mathcal{Z})$. Its particular values are:

$$\begin{aligned}
\mathrm{dn}\,(0, m) &= 1, & \mathrm{dn}\,(\pm K, m) &= \sqrt{m'}, \\
\mathrm{dn}\,(\pm K + iK', m) &= 0, & \mathrm{dn}\,(\pm iK', m) &= \infty.
\end{aligned}$$

The function $w = \mathrm{dn}\,(z, m)$ conformally maps the basic rectangle onto the half-plane Re $w > 0$ with the cut along the segment $[0, 1]$. The mapping of the Cartesian net in the basic rectangle realized by the function $w = \mathrm{dn}\,(z, m)$ with $m = 0.7$ is shown in Fig. 1.139, b by solid lines.

The function $w = \mathrm{dn}\,(z, m)$ is univalent in the doubled rectangle $-K < x < K$, $0 < y < 2K'$ and conformally maps it onto the plane w with the cut along the segment $[-1, 1]$. The mapping of the Cartesian net in the doubled basic rectangle is also shown in Fig. 1.139, b.

1.14   Conformal Mapping of Doubly Connected Domains

Any doubly connected domain with nondegenerate boundaries can be conformally mapped onto some annulus $1 < |z| < R$ (see Subsection 1.6.2). The number R is called the *modulus of the doubly connected domain* $(R > 1)$.

The annulus $1 < |z| < R$ can be considered as an adequate canonical domain for all doubly connected domains with the given modulus R. The simplest domain different from an annulus is the domains bounded by two eccentric non-intersecting circumferences (or a circumference and a straight line). Such domains are conformally mapped onto an annulus by linear-fractional function.

Let the boundary of the doubly connected domain G in the plane w be two circumferences: $|w| = 1$ and $|w - L| = R_0$, where L is the distance between the centers of the circumferences. Consider first the case $R_0 > L + 1$ when the domain G is an eccentric annulus between two circumferences, one of which is lying inside the other: $|w| > 1$, $w - L < R_0$. Designate as c and d the intersections of the greater circumference $|w - L| = R_0$ with the real axis: $c = L - R_0$, $d = L + R_0$ (here $c < -1$, $d > 1$).

With the help of the linear-fractional transformation with the real coefficients it is possible to map the domain G onto some circular annulus $1 < |z| < R$ so that the points $w_1 = c$, $w_2 = -1$, $w_3 = 1$, $w_4 = d$ would be mapped into the points $z_1 = -R$, $z_2 = -1$, $z_3 = 1$, $z_4 = R$. The modulus of the doubly connected domain R can be determined by the invariance condition of the *anharmonic ratio* of four points:

$$D = \frac{z_3 - z_1}{z_3 - z_2} : \frac{z_4 - z_1}{z_4 - z_2}.$$

For four of the points $(c, -1, 1, d)$ the anharmonic ratio is equal to

$$D = \frac{1 - c}{2} \cdot \frac{d + 1}{d - c}, \tag{1.14.1}$$

and $D > 1$. For four of the points $(-R, -1, 1, R)$ the anharmonic ratio is equal to

$$\frac{(1 + R)^2}{4R}.$$

The quadratic equation

$$\frac{(1 + R)^2}{4R} = D$$

has two real roots for $D > 1$, one of them being greater than 1:

$$R = \left(\sqrt{D} + \sqrt{D - 1}\right)^2. \tag{1.14.2}$$

The sought for linear-fractional mapping of the annulus can be found from the condition of equality of anharmonic ratios for two groups of four points $(-R, -1, 1, z)$ and $(c, -1, 1, w)$. Designating the anharmonic ratio of these points as t we have

$$t = \frac{R + 1}{2} \cdot \frac{z + 1}{z + R}, \tag{1.14.3}$$

$$w = \frac{2tc + 1 - c}{2t - 1 + c}. \tag{1.14.4}$$

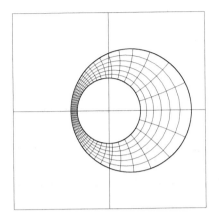 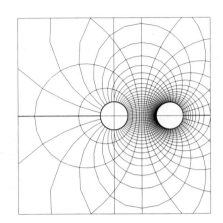

FIGURE 1.140

Thus the function mapping the annulus $1 < |z| < R$ onto the eccentric annulus is the composition of two linear-fractional functions (1.14.3) and (1.14.4), which depends on two parameters c and R. Practically, it is more convenient not to calculate the parameter R, but to fit it in order to obtain the necessary shape of a domain.

The mapping of the polar net in the annulus $1 < |z| < R$ carried out by the function (1.14.3)–(1.14.4) with $R = 1.5$, $c = -1.2$ is shown in Fig. 1.140, a.

In the case $L > R_0 + 1$ the domain G is the plane w with two circles deleted:

$$|w| > 1, \quad |w - L| > R_0.$$

In this case take $c = L + R_0$, $d = L - R_0$ ($1 < d < c$ holds true). In these designations the formulae (1.14.1)–(1.14.2) for modulus calculation and (1.14.3)–(1.14.4) for the calculation of mapping function stay unchanged. The mapping of the polar net in the annulus $1 < |z| < R$ realized by the function (1.14.3)–(1.14.4) with $c = 5$, $R = 15$ is shown in Fig. 1.140, b. Note that the image of the polar net is the bipolar one.

Each function analytic and univalent in the circular annulus maps it conformally onto some doubly connected domain. It is possible to construct conformal mappings of some interesting domains by elementary functions, that are univalent in a circle (or in the exterior of a circle).

For example, the circular annulus $r_1 < |z| < r_2$, where $0 < r_1 < r_2 \leq 1$, is mapped onto:

(a) the domain between two *Pascal's limaçons* (Subsection 1.10.4) by the function $w = (1 + z)^2$,

(b) the domain between two *shortened epitrochoids* (Subsection 1.10.6) by the function $w = z + z^n/n$,

(c) the domain between two confocal elliptic *Booth's lemniscatas* (Subsection 1.10.9) by the function $w = 2z/(z^2 + 1)$.

Analogously the circular annulus $r_1 < |z| < r_2$ (where $1 \leq r_1 < r_2$) is mapped onto:

(a) the domain between two confocal ellipses (Subsection 1.7.4) by the Zhukovskii function $w = 1/2\,(z + 1/z)$,

(b) the domain between two confocal *single-contour Cassini's ovals* (Subsection 1.10.2) by the function $w = z\sqrt{1 + 1/z^2}$,

(c) the domain between two *shortened hypocycloids* (Subsection 1.10.7) by the function $w = z + 1/(nz^n)$.

To construct a conformal mapping of a doubly connected domain elliptical functions are used rather effectively. Using them a circular annulus can be mapped onto a plane with two cuts along segments of the same straight line. It can be realized as follows.

Let a doubly connected domain G be the annulus $1/R < |z| < R$. Making a cut along a segment of the real axis $-R < x < -1/R$ it is possible to map the obtained singly connected domain by the function $\zeta = \ln z$ onto a rectangle. The rectangle in the plane w is $-b < \mathrm{Re}\,\zeta < b$, $-\pi < \mathrm{Im}\,\zeta < \pi$, where $b = \ln R$ and the lips of the cut are mapped onto the horizontal sides of the rectangle $\mathrm{Im}\,\zeta = \pm\pi$. The rectangle is mapped onto the plane t with the cuts along the rays $(-\infty, -1)$ and $(1, \infty)$ by the function $t = \mathrm{sn}\,(\zeta K/b, m)$ with m being the root of the equation

$$\frac{K(1 - m)}{K(m)} = \frac{\pi}{b}$$

(Subsection 1.13.5). Due to the periodicity of the function $\mathrm{sn}\,(z, m)$ the following relation holds true

$$\mathrm{sn}\,(x + iK', m) = \mathrm{sn}\,(x - iK', m),$$

i.e., the points of the upper and the lower sides of the rectangle with the same abscissas are mapped into the same point.

Thus the circular annulus with the cut along the segment of the real axis $(-R, -1/R)$ is mapped onto the plane t with cuts along the rays $(-\infty, -1)$ and $(1, \infty)$ by the function

$$\zeta = \ln z, \quad t = \mathrm{sn}\,(\zeta K/b, m). \tag{1.14.5}$$

The points positioned at the opposite lips of the cut $-R < \mathrm{Re}\,z < -1/R$ are mapped into the points lying at the opposite lips of the cuts along the rays $(-\infty, -1/\sqrt{m})$ and $(1/\sqrt{m}, \infty)$. Therefore these cuts can be ignored. The doubly connected annulus $1/R < |z| < R$ can be considered to be mapped conformally by the function (1.14.5) onto the doubly connected domain — the plane t with the cuts along segments of the real axis $(-1/\sqrt{m}, -1)$ and $(1, 1/\sqrt{m})$. The rectangle on the plane ξ of the intermediate variable can also be considered as the doubly connected domain, namely, as the rectangle with glued pair of opposite horizontal sides. The Cartesian net in the rectangle is mapped on one hand onto the polar net in the circular annulus and on the other hand — onto the net of lines on the plane t shown in Fig. 1.125, a. This net can be interpreted as the net of force lines and equipotentials of the electric field created by two equal conductive coplanar strips carrying charges of opposite signs (see Subsection 2.2.4 for details).

The mapping realized by the function $w = \mathrm{cn}\,(z, m)$ (Fig. 1.138, b) can also be interpreted as the figure of the electric field in the doubly connected domain bounded by two semi-infinite cuts along the rays of the real axis $(-\infty, -1)$ and $(1, \infty)$ and a finite cut along the segment of the imaginary axis $(-i\sqrt{m'/m}, i\sqrt{m'/m})$.

A number of other mappings of doubly connected domains can be built by the conformal mapping of domains shown in Fig. 1.125, a. For example, a wide set of doubly connected domains can be obtained by the linear-fractional transformations.

Example 1.51
The plane t with two symmetrical cuts along the segments $(-1/\sqrt{m}, -1)$ and $(1, 1/\sqrt{m})$ can be mapped onto the plane w with two cuts along arbitrary segments of the real axis (a, b) and (c, d) by the linear-fractional transformation with real coefficients. In particular if $a = \infty$, $b = 0$, $c = 1$, such a mapping is realized by the function

$$w = \frac{p+1}{2} \cdot \frac{t+1}{tp+1}, \quad p = \sqrt{m}.$$

Therefore, the sequence of transformations

$$p = \sqrt{m},$$
$$t = \mathrm{sn}\,(z, m), \tag{1.14.6}$$
$$w = \frac{p+1}{2} \cdot \frac{t+1}{tp+1}$$

maps the rectangle $-K < x < K$, $-K' < y < K'$ onto the plane w with two cuts: one along the ray $(-\infty, 0)$ and the other along the segment $(1, d)$, where $d = (p+1)^2/(4p)$.

The mapping realized by the function (1.14.6) with $m = 0.01$ is shown in Fig. 1.141, a.

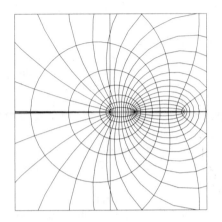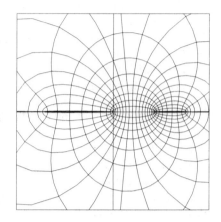

FIGURE 1.141

Example 1.52

The mapping of the rectangle $-K < x < K$, $-K' < y < K'$ onto the plane with the cuts along the segments $(-1, 0)$ and (c, d) is realized by the sequence of transformations

$$p = \sqrt{m}\,,$$

$$t = \operatorname{sn}(z, m)\,,$$

$$D = \frac{p+1}{2} \cdot \frac{t+1}{tp+1}\,,$$

$$w = \frac{cD}{c+1-cD}\,.$$

The mapping realized by the function (1.14.6) with $m = 0.1$, $c = 0.6$ is shown in Fig. 1.141, b.

Example 1.53

The function $w = (1+t)/(1-t)$, where $t = \operatorname{cn}(z, m)$, maps the rectangle $-K < x < K$, $0 < y < 2K'$ onto the doubly connected domain shown in Fig. 1.142. This mapping is built with $m = 0.6$.

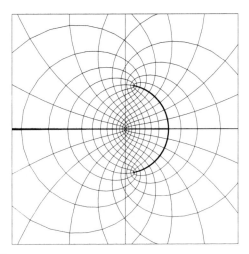

FIGURE 1.142

Example 1.54

The function $w = (1+t)/(1-t)$, where $t = \operatorname{dn}(z, m)$, conformally maps the basic rectangle $-K < x < K$, $0 < y < K'$ onto the doubly connected domain shown in Fig. 1.143. This mapping is built with $m = 0.9$.

If a doubly connected domain bounded by two cuts along segments of the real axis is subjected to the linear-fractional transformation with complex coefficients, a doubly connected domain bounded by two arcs of the same circumference will be obtained.

Example 1.55

The function $w = (1+it)/(1-it)$, $t = \sqrt[4]{m} \operatorname{sn}(z, m)$ conformally maps the rectangle $-K < x < K$, $-K' < y < K'$ onto the doubly connected domain shown in Fig. 1.144. This mapping is built with $m = 0.01$.

With the help of the function $w = \sqrt{t^2 - 1}$ it is possible to map a plane with two cuts along segments of the same straight line onto a doubly connected domain bounded by reciprocally-perpendicular segments.

Example 1.56

The mapping of the Cartesian net in the rectangle $-K < x < K$, $-K' < y < K'$ realized by the composition of functions

$$s = \operatorname{sn}(z, m),$$

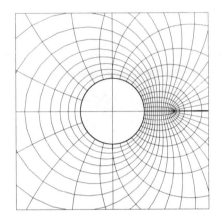

FIGURE 1.143

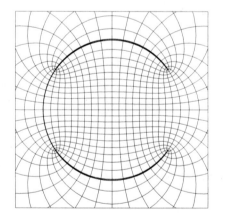

FIGURE 1.144

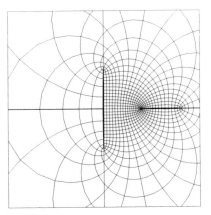 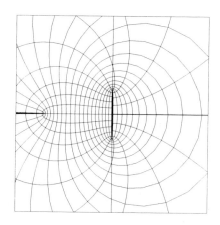

FIGURE 1.145

$$D = \frac{p+1}{2} \cdot \frac{s+1}{sp+1},$$

$$t = \frac{a+D}{a-D},$$

$$w = \sqrt{t-1}\,\sqrt{t+1}$$

is presented in Fig. 1.145, a with $p = \sqrt{m}$, $m = 0.02$, $a = 6$.

Example 1.57

An analogous mapping of a doubly connected domain bounded by the ray and the segment perpendicular to it is shown in Fig. 1.145, b. It is realized by the composition of functions

$$s = \operatorname{sn}^2(z, m),$$

$$t = \frac{1+m}{1-m}(s-1) - s,$$

$$w = \sqrt{t-1}\,\sqrt{t+1}$$

with $m = 0.5$.

A doubly connected domain bounded by a circumference and a segment is obtained under the conformal mapping of the plane with two cuts by the inverse Zhukovskii function.

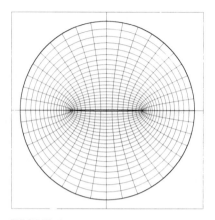 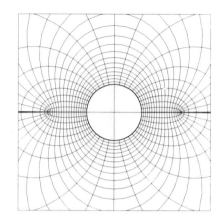

FIGURE 1.146

Example 1.58

The mapping of the Cartesian net in the rectangle $-K < x < K$, $-K' < y < K'$ carried out by the function

$$w = t - \sqrt{t+1}\,\sqrt{t-1},$$

where $t = \mathrm{sn}\,(z,m)$, $m = 0.5$ is shown in Fig. 1.146, a.

The mapping realized by the other branch of the inverse Zhukovskii function $w = t + \sqrt{t+1}\,\sqrt{t-1}$ is shown in Fig. 1.146, b.

By making the conformal mapping of a plane with two cuts by the function $w = \arcsin t$, the doubly connected domain is obtained, being an infinite strip with a cut along some segment.

Example 1.59

The mapping of the Cartesian net in the rectangle $-K < x < K$, $-K' < y < K'$ realized by the function

$$t = \sqrt{m}\,\mathrm{sn}\,(z,m),\quad w = i\,\arcsin t$$

with $m = 0.5$ is shown in Fig. 1.147, a.

The conformal mapping of a half-strip with a cut along a segment can be built analogously.

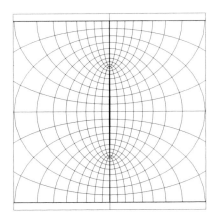 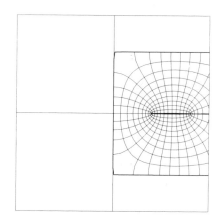

FIGURE 1.147

Example 1.60
The mapping of the Cartesian net in the rectangle $0 < x < K$, $-K' < y < K'$ realized by the function

$$s = \operatorname{sn}(z, m),$$

$$t = \frac{is}{p},$$

$$w = -i \arcsin t$$

with $m = 0.1$, $p = 0.865$ is shown in Fig. 1.147, b.

Conformal mappings of doubly connected domains bounded by cuts along segments of straight lines and conic curves are constructed in the following three examples.

Example 1.61
The composition of functions

$$s = A \operatorname{sn}(z, m), \quad (A > 0)$$

$$w = (1 + s)^2 - 2$$

maps the rectangle $0 < x < K$, $-K' < y < K'$ onto the doubly connected domain, being the exterior of a parabola with the cut along a segment of the real axis. The mapping realized by the function with $m = 0.5$, $A = 1$ is presented in Fig. 1.148, a.

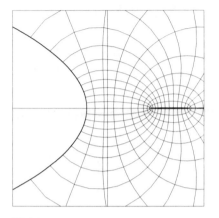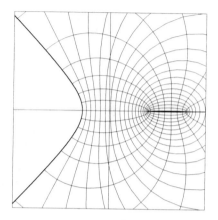

FIGURE 1.148

Example 1.62

The mapping of the same rectangle onto the exterior of a hyperbola with
the cut along a segment of the real axis is presented in Fig. 1.148, b. The
mapping is created with the help of the function

$$s = \operatorname{sn}(z, m),$$

$$t = 2p \arccos s,$$

$$w = \cos t$$

with $m = 0.5$, $p = 0.75$.

Example 1.63

The function

$$d = \operatorname{dn}(z, m),$$

$$t = A\frac{1+d}{1-d},$$

$$w = t - \frac{1}{t}$$

maps the basic rectangle $-K < x < K$, $0 < y < K'$ onto the doubly
connected domain bounded by the ellipse and the infinite cut along the ray
of the real axis. The mapping carried out by this function with $m = 0.9$,
$A = 1.2$ is shown in Fig. 1.149.

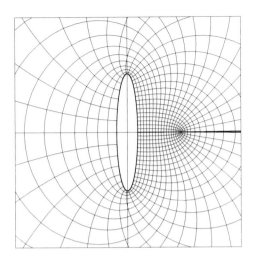

FIGURE 1.149

Example 1.64
The composition of functions

$$s = H \operatorname{sn}(z, m),$$

$$v = \sqrt{s - H} \sqrt{s + H},$$

$$u = v + \sqrt{v - 1} \sqrt{v + 1},$$

$$t = R u,$$

$$w = t \sqrt{1 + 1/t^2},$$

where $R = \sqrt{1 - m}/(1 + \sqrt{m})$, $H = (1/R - R)/2$, maps the rectangle $-K < x < K$, $-K' < y < K'$ onto the doubly connected domain being the exterior of the *two-contour Cassini's oval*. The mapping realized by the function with $m = 0.001$ is shown in Fig. 1.150

Interesting doubly connected domains can be obtained mapping a plane with two cuts along segments of the real axis (Fig. 1.137, b) by elliptic integrals.

Example 1.65
The function $w = F(t, p)$, $t = \sqrt{m} \operatorname{sn}(z, m)$ maps the rectangle $-K < x < K$, $-K' < y < K'$ onto the rectangle with the cut along the segment of the real axis. The mapping realized by this function with $m = 0.5$, $p = 0.8925$ is shown in Fig. 1.151.

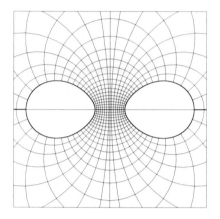

FIGURE 1.150

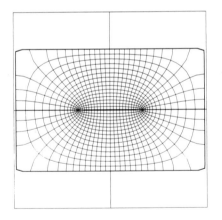

FIGURE 1.151

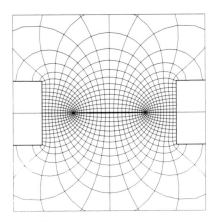 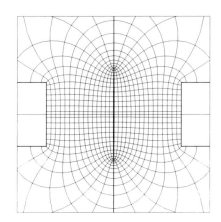

FIGURE 1.152

Example 1.66
The function $w = E(t,p)$, $t = \sqrt{m}\,\text{sn}\,(z,m)$ maps the rectangle $-K < x < K$, $-K' < y < K'$ onto the doubly connected domain shown in Fig. 1.152, a. This mapping is built with $m = 0.5$, $p = 0.4$.

Example 1.67
The function $w = E(c,p)$, $c = \text{cn}\,(z,m)$ maps the rectangle $-K < x < K$, $0 < y < 2K'$ onto the doubly connected domain shown in Fig. 1.152, b. This mapping is built with $m = 0.5$, $p = 0.4$.

Example 1.68
By substituting $\text{sn}\,(z,m)$ instead of z in the function (1.13.17), the so-called *Z-Jacobian function* will be obtained:

$$s = \text{sn}\,(z,m),$$

$$w = E(s,m) - \frac{E}{K}\,z. \tag{1.14.7}$$

The function (1.14.7) maps the basic rectangle $-K < x < K$, $0 < y < K'$ onto the doubly connected domain, being the half-plane with the cut along a segment. The mapping realized by the function (1.14.7) with $m = 0.9$ is shown in Fig. 1.153, a.

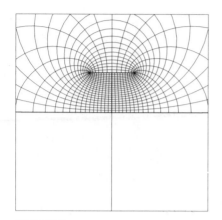 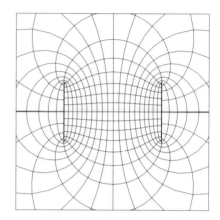

FIGURE 1.153

Example 1.69

The function $w = E(s,m) - Az$, where $s = \operatorname{sn}(z,m)$, $A = 1 - E'/K'$, $m = 0.1$ maps the rectangle $-K < x < K$, $-K' < y < K'$ onto the doubly connected domain with the boundary consisting of two segments of parallel straight lines. The mapping of the Cartesian net in the rectangle realized by the function is shown in Fig. 1.153, b.

2

Physics: Applications of Conformal Mappings

Conformal mappings are widely used in mathematical physics when solving problems of hydrodynamics,[109-126] of the theory of filtration,[127-129] of the theory of electromagnetic field,[132-146] of the theory of heat conductivity,[163] and of the theory of elasticity.[147-162] The main field of conformal mappings applications is the calculation of plane harmonic vector fields.

2.1 Insight into Plane Harmonic Vector Field

A vector field $\mathbf{A}$ is called the *plane* (or plane-parallel) one, if at all the points of the space the vector $\mathbf{A}$ is parallel to some plane S and if it has the same value and direction at all the points of any straight line perpendicular to the plane S.

Assume that the plane S coincides with the plane x, y of the Cartesian coordinate system x, y, Z, and $\mathbf{i}, \mathbf{j}, \mathbf{k}$ are unit vectors directed along axis x, y, Z, correspondingly. In this coordinate system the plane field is

$$\mathbf{A} = P(x,y)\,\mathbf{i} + Q(x,y)\,\mathbf{j}, \qquad (2.1.1)$$

where $P(x,y)$ and $Q(x,y)$ are components of the vector $\mathbf{A}$ onto coordinate axis.

The vector field $\mathbf{B}$ directed along the normal to the plane S and having the same value at all the points of any straight line perpendicular to S is called the *anti-plane* one. In the x, y, Z coordinate system the anti-plane field is

$$\mathbf{B} = B(x,y)\,\mathbf{k}.$$

Note that the curl of the plane vector field is the anti-plane field:

$$[\nabla \times \mathbf{A}] = \begin{vmatrix} \mathbf{i} & \mathbf{j} & \mathbf{k} \\ \partial/\partial x & \partial/\partial y & \partial/\partial z \\ P & Q & 0 \end{vmatrix} = \mathbf{k}\left(\frac{\partial Q}{\partial x} - \frac{\partial P}{\partial y}\right). \qquad (2.1.2)$$

In turn the curl of the anti-plane field is a plane field:

$$[\nabla \times \mathbf{B}] = \mathbf{i}\frac{\partial B}{\partial y} - \mathbf{j}\frac{\partial B}{\partial x}.$$

A vector field $\mathbf{A}$ is called *potential* if $[\nabla \times \mathbf{A}] = 0$.
A vector field $\mathbf{A}$ is called *solenoidal* if $(\nabla \cdot \mathbf{A}) = 0$.
A vector field $\mathbf{A}$ being simultaneously potential and solenoidal is called *harmonic*.
The plane field (2.1.1) is potential if

$$\frac{\partial Q}{\partial x} - \frac{\partial P}{\partial y} = 0. \qquad (2.1.3)$$

The plane field (2.1.1) is solenoidal if

$$\frac{\partial P}{\partial x} + \frac{\partial Q}{\partial y} = 0. \qquad (2.1.4)$$

The conditions of harmonicity of the plane vector field (2.1.3–2.1.4) are Cauchy-Riemann conditions for the function $F(z) = P(x,y) - iQ(x,y)$, where i is the imaginary unit (not to be confused with the vector $\mathbf{i}$), $z = x + iy$ (not to be confused with the Cartesian coordinate Z).
A vector of the plane xy

$$\mathbf{A}^* = [\mathbf{k} \times \mathbf{A}] = -Q(x,y)\,\mathbf{i} + P(x,y)\,\mathbf{j}$$

is the vector $\mathbf{A}$ turned on the angle $\pi/2$ in the counter-clockwise direction. For the plane vector field $\mathbf{A}^*$

$$[\nabla \times \mathbf{A}^*] = \mathbf{k}\left(\frac{\partial P}{\partial x} + \frac{\partial Q}{\partial y}\right) = \mathbf{k}\,(\nabla \cdot \mathbf{A}),$$

$$(\nabla \cdot \mathbf{A}^*) = -\frac{\partial Q}{\partial x} + \frac{\partial P}{\partial y} = -(\mathbf{k}\,[\nabla \times \mathbf{A}]).$$

If the plane vector field $\mathbf{A}$ is harmonic the field $\mathbf{A}^*$ is also harmonic. It is called *harmonically conjugate* to the vector field $\mathbf{A}$.
Let the plane vector field $\mathbf{A} = P(x,y)\,\mathbf{i} + Q(x,y)\,\mathbf{j}$ be given in the plane domain G; L be a smooth curve in the domain, and $d\mathbf{r} = \mathbf{i}\,dx + \mathbf{j}\,dy$ be an arc element of the curve L. The curvilinear integral

$$\Gamma = \int_L (P\,dx + Q\,dy) = \int_L (\mathbf{A} \cdot d\mathbf{r}) \qquad (2.1.5)$$

is called the *work of the vector* **A** along the curve L. When the curve L is closed the integral (2.1.5) is called a *circulation* of vector **A**.

A curvilinear integral

$$N = \int_L (-Q\,dx + P\,dy) = \int_L (\mathbf{A}^* \cdot d\mathbf{r}) \qquad (2.1.6)$$

is called a *flux of the vector* **A** across the curve L.

If the domain G is singly connected and the field **A** is harmonic the integrals (2.1.5) and (2.1.6) vanish around any closed curve in accordance with the known theorem of vector analysis. Integrals along non-closed curves L connecting two fixed points $M_0(x_0, y_0)$ and $M_1(x_1, y_1)$ do not depend on the shape of the curve L.

2.2 Plane Harmonic Vector Fields in Physics

2.2.1 Velocity field of steady-state flow of ideal liquid

A velocity field **v** of the motion of incompressible liquid satisfies the condition

$$(\nabla \cdot \mathbf{v}) = 0.$$

We consider only potential flows of ideal (non-viscous) liquid with $[\nabla \times \mathbf{v}] = 0$. Therefore, the velocity field **v** is harmonic in the potential flow of noncompressible liquid.

An example of a plane harmonic field is the velocity field of ideal liquid in potential streaming around the infinite cylinder or in the potential flow of liquid in the curvilinear canal with constant depth. On the canal wall or on the surface of a motionless cylinder the field **v** satisfies the condition $v_n = 0$, where $v_n = (\mathbf{n} \cdot \mathbf{v})$ is the normal component of the velocity of the liquid.

2.2.2 Velocity field of liquid in the steady-state filtration

The velocity of liquid moving in the porous medium (for example, underground water in the soil) is described by the *Darcy law*:

$$\mathbf{v} = -D\,\nabla p, \qquad (2.2.1)$$

where p is the pressure, D is the constant characterizing the porous medium. The formula (2.2.1) yields that the velocity field of the liquid in the homogeneous porous medium is potential: $[\nabla \times \mathbf{v}] = 0$. If the liquid and the porous medium are incompressible, then $(\nabla \cdot \mathbf{v}) = 0$. Thus the

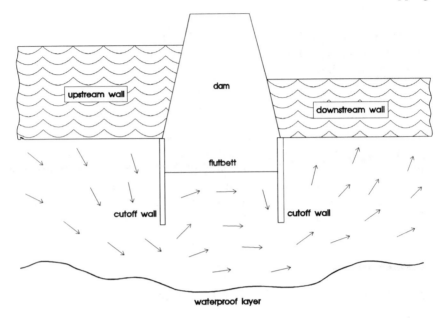

FIGURE 2.1

velocity field of incompressible liquid filtered in the homogeneous incompressible medium is harmonic.

An example of a plane harmonic field in the theory of filtration is the velocity field of liquid in homogeneous ground lying under a long dam built on the wide channel with constant depth (Fig. 2.1).

At the waterproof boundary of the porous medium (flutbett, waterproof layer) the velocity $\mathbf{v}$ satisfies the condition $v_n = 0$, where $v_n = (\mathbf{n} \cdot \mathbf{v})$ is the normal component of the velocity of the liquid.

2.2.3   Stationary problems of heat conductivity

A vector of the density of the heat flux $\mathbf{q}$ in an inhomogeneously heated body is defined by the *Fourier law*:

$$\mathbf{q} = -\lambda \nabla u, \qquad (2.2.2)$$

where u is the temperature and λ is the thermal conductivity. Equation (2.2.2) yields for $\lambda = \text{const}$ that $[\nabla \times \mathbf{q}] = 0$, i.e., in the homogeneous medium the vector of heat flux (2.2.2) is a potential vector field. If there are no volume heat sources inside the medium, the steady-state heat flux is a solenoidal field: $(\nabla \cdot \mathbf{q}) = 0$. A steady-state heat flux in an homogeneous plate is an example of a plane harmonic vector field.

2.2.4 Problems of electrostatics

An electromagnetic field **E**, **H** in the vacuum is described by the system of *Maxwell's equations**

$$[\nabla \times \mathbf{E}] = -\frac{1}{c}\frac{\partial \mathbf{H}}{\partial t},$$

$$[\nabla \times \mathbf{H}] = \frac{1}{c}\frac{\partial \mathbf{E}}{\partial t} + \frac{4\pi}{c}\mathbf{J}, \qquad (2.2.3)$$

$$(\nabla \cdot \mathbf{E}) = 4\pi\rho,$$

$$(\nabla \cdot \mathbf{H}) = 0,$$

where c is the light speed, **J** is the density of electric current, and ρ is the density of the electric charge.

The electrostatic field in an homogeneous medium free from electric charges satisfies the equations

$$[\nabla \times \mathbf{E}] = 0, \quad (\nabla \cdot \mathbf{E}) = 0,$$

i.e., is the harmonic one.

A field created by an infinite charged conductive cylinder with an arbitrary cross-section is an example of a plane electric field.

2.2.5 Problems of magnetostatics

The Maxwell's equations (2.2.3) yield that steady magnetic field **H** is harmonic in a homogeneous medium free of electric currents:

$$[\nabla \times \mathbf{H}] = 0, \quad (\nabla \cdot \mathbf{H}) = 0.$$

The magnetic field of a current flowing along an infinite cylinder with an arbitrary cross-section is an example of the plane magnetic field. Outside the cylinder the field **H** is the plane harmonic vector field. If the cylinder is superconductive, the magnetic field inside the cylinder vanishes (the so-called *Meissner's effect*). On the surface of a superconductive cylinder the magnetic field satisfies the condition $(\mathbf{n} \cdot \mathbf{H}) = 0$, i.e., the field **H** is directed along the tangent to the cylinder surface.

Another example of a plane magnetic field is the field in a gap between magnet poles being two infinite cylindrical surfaces with infinite directrix. If the poles are made of the magnetic material with infinitely large magnetic permittivity, the field **H** satisfies the condition $[\mathbf{n} \times \mathbf{H}] = 0$ on the surface of the pole, i.e., the magnetic field is directed along the normal to the pole surface.

---

*We use Gaussian units. The usage of other units in equations (2.2.3) changes only its coefficients. Maxwell's equations for the field in the homogeneous medium other than vacuum also differ from the equations (2.2.3) only by the constant coefficients.

2.2.6 Steady-state electric current in homogeneous conductive medium

The density of electric current in the medium with electrical conductivity σ is defined by the *Ohm's law*: $\mathbf{J} = \sigma\mathbf{E}$. Maxwell's equations (2.2.3) yield that in an homogeneous medium ($\sigma = \text{const}$) the density of electric current $\mathbf{J}$ is a harmonic vector field

$$[\nabla\times\mathbf{J}] = 0, \quad (\nabla\cdot\mathbf{J}) = 0.$$

The distribution of the electric current created by electrodes placed at the edge of an homogeneous conductive plate is an example of the plane harmonic field of the current.

2.2.7 Transversal electromagnetic waves in waveguiding systems

Certain transmitting lines of electromagnetic waves infinite along Z-axis give rise to electromagnetic waves described by the equations

$$\mathbf{E} = \mathbf{A}(x,y)\,f(Z - ct)\,, \quad \mathbf{H} = \mathbf{A}^*(x,y)\,f(Z - ct)\,. \tag{2.2.4}$$

Here $\mathbf{A}$ is a plane harmonic field, $\mathbf{A}^*$ is a harmonically conjugate field, and f is an arbitrary differentiable function of a single variable.[†] It is easy to check that vectors $\mathbf{E}$ and $\mathbf{H}$ given by (2.2.4) satisfy the set of Maxwell's equations (2.2.3). Indeed, well-known formulae of vector analysis yield

$$[\nabla\times\mathbf{E}] = [\nabla f \times \mathbf{A}] = [\mathbf{k}\times\mathbf{A}]\,f' = \mathbf{A}^*f'(Z - ct)\,,$$

$$[\nabla\times\mathbf{H}] = [\nabla f \times \mathbf{A}^*] = [\mathbf{k}\times\mathbf{A}^*]\,f' = -\mathbf{A}f'(Z - ct)\,.$$

Further we have

$$\frac{\partial\mathbf{E}}{\partial t} = -cf'(Z - ct)\,\mathbf{A}\,, \quad \frac{\partial\mathbf{H}}{\partial t} = -cf'(Z - ct)\,\mathbf{A}^*$$

and finally

$$(\nabla\cdot\mathbf{E}) = f\,(\nabla\cdot\mathbf{A}) = 0\,, \quad (\nabla\cdot\mathbf{H}) = f\,(\nabla\cdot\mathbf{A}^*) = 0\,.$$

Thus the electromagnetic wave (2.2.4) is the solution of equations (2.2.3) with $\rho = 0$, $\mathbf{J} = 0$. It is called *transversal electromagnetic wave* (or TEM-wave), propagating along Z-axis. On the ideally conductive surface (the wall of a waveguiding system) the tangent component of the field $\mathbf{E}$ and the normal component of the field $\mathbf{H}$ should vanish:

$$[\mathbf{n}\times\mathbf{A}] = (\mathbf{n}\cdot\mathbf{A}^*) = 0\,.$$

[†]Note that the fields $\mathbf{E}$ and $\mathbf{H}$ are neither plane nor harmonic ones.

2.3 Complex Potential

Let $\mathbf{A} = P(x,y)\mathbf{i} + Q(x,y)\mathbf{j}$ be a plane harmonic vector field given in singly connected plane domain G. According to the known theorem of vector analysis there exists a function $U(x,y)$ in the domain G such that

$$\mathbf{A} = \frac{\partial U}{\partial x}\mathbf{i} + \frac{\partial U}{\partial y}\mathbf{j} = \nabla U. \qquad (2.3.1)$$

The function $U(x,y)$ is called *scalar potential* (or simply *potential*) of the vector field $\mathbf{A}$. It is defined to within an additive constant. Due to the condition $\nabla \cdot \mathbf{A} = 0$ the function $U(x,y)$ is harmonic: $\nabla^2 U = 0$.

In the Section 1.3 it was mentioned that for any function $U(x,y)$ harmonic in a singly connected domain G there exists a harmonically conjugate function $V(x,y)$ such that the complex function

$$f(x,y) = U(x,y) + iV(x,y) \qquad (2.3.2)$$

is analytic in the domain G. The function $V(x,y)$ is also defined to within an additive constant.

Due to the Cauchy-Riemann conditions for the function f the field $\mathbf{A}$ (2.3.1) can be presented as

$$\mathbf{A} = \frac{\partial V}{\partial y}\mathbf{i} - \frac{\partial V}{\partial x}\mathbf{j} = \nabla \times (\mathbf{k}V).$$

The function $V(x,y)$ is called a *vector potential** of the harmonic vector field $\mathbf{A}$.

The analytic function (2.3.2) is called a *complex potential* of the vector field $\mathbf{A}$.

Let the function of a complex variable

$$a(z) = P(x,y) + iQ(x,y),$$

where $z = x + iy$, be brought into correspondence to the vector field $\mathbf{A} = P\mathbf{i} + Q\mathbf{j}$. Unlike the function $a(z)$ the function $\overline{a(z)}$ is analytic and

$$\overline{a(z)} = \frac{\partial U}{\partial x} + i\frac{\partial V}{\partial x} = f'(z).$$

Therefore, the function $a(z)$ representing the plane harmonic vector field $\mathbf{A}$ is rather simply expressed through the complex potential $f(z)$:

$$a(z) = \overline{f'(z)}. \qquad (2.3.3)$$

---

*In hydrodynamics the function V is usually called the *stream function*, in electrodynamics — the *force function*.

The work (2.1.5) and the flux (2.1.6) of a harmonic vector field are also rather simply expressed through a complex potential. Let L be a curve connecting two points of a singly connected domain $z_0 = x_0 + iy_0$ and $z_1 = x_1 + iy_1$. The work of the vector field $\mathbf{A}$ along the curve L (2.1.5) and the flux of the vector across the curve L (2.1.6) are expressed through the complex potential as:

$$\Gamma + iN = \int_{z_0}^{z_1} \overline{a(z)}\, dz = f(z_1) - f(z_0),$$

that is

$$\Gamma = \mathrm{Re}\,[f(z_1) - f(z_0)], \quad N = \mathrm{Im}\,[f(z_1) - f(z_0)].$$

We will use the complex potential for the visualization of plane harmonic vector fields. Any plane potential field can be depicted on the plane as a net of equipotentials and vector lines of the field.[†] A harmonic plane vector field is visualized as an isothermic net of the complex potential. Critical points of the complex potential (i.e., points where $f'(z) = 0$) are singular points of the vector field $\mathbf{A}$, because $\mathbf{A} = 0$ at those points and the direction of the field is undefined.

If the plane harmonic field $\mathbf{A}$ has the complex potential $f(z)$, the harmonically conjugate field $\mathbf{A}^*$ has the complex potential $-if(z)$. Equipotentials of the vector field $\mathbf{A}$ are the vector lines of harmonically conjugate field A^* and vice versa. Thus, having built the isothermic net of one analytic function $f(z)$ we obtain a net of equipotentials and vector lines of two harmonically conjugate vector fields $\mathbf{A}$ and $\mathbf{A}^*$.

Note at the conclusion that the concept of a complex potential can also be introduced for the harmonic vector field given at a multiply connected domain. However in this case the complex potential can be a multiply-valued (infinite-valued) function with branches differing one from another by the additive constants, called cyclic constants.

2.4   Boundary-Value Problems for Harmonic Functions

A problem of determination of the harmonic vector field is usually formulated as a boundary-value problem for the real-valued harmonic function

---

[†] *Equipotentials* of the potential field $\mathbf{A}$ are level lines of a scalar potential, i.e., the lines defined by the equation $U(x,y) = C$. *Vector line* of the vector field $\mathbf{A}$ is the line with the tangent at any point being collinear to the direction of the vector $\mathbf{A}$ at this point (in hydrodynamics the vector lines are called *streamlines*, in electrodynamics — *force lines*. For the harmonic vector field vector lines are level lines of the vector potential: $V(x,y) = C$.

$U(x, y)$ that is the potential of the field. The function harmonic in the domain function can be found according to conditions at the boundary of the domain and (in the case of the infinite domain) — according to conditions at the point at infinity. Let a smooth curve L be a part of the boundary of the domain of harmonicity of the function $U(x, y)$. In this handbook two types of boundary conditions are considered:

1. Boundary conditions of the first type, when on the curve L the value of $U(x, y)$ is given:

$$U|_L = f(x, y) \,; \qquad (2.4.1)$$

2. Boundary conditions of the second type, when on the curve L the normal derivative is given:

$$\left. \frac{\partial U}{\partial n} \right|_L = g(x, y) \,, \qquad (2.4.2)$$

where n is the normal to the curve L.

Let G be a plane domain with the boundary ∂G. If for the whole the boundary ∂G boundary conditions of the first type (2.4.1) are given for the harmonic function $U(x, y)$, the problem of determination of the function U is called the *Dirichlet problem*. The problem of determination of the harmonic function according to the given boundary condition of the second type (2.4.2) is called the *Neumann problem*. If on the part of the boundary of the domain the boundary condition of the first type is given, while on the other part of the boundary — that of the second type is given, such a boundary-value problem is called the *mixed* one.

In the course of mathematical physics it is proved that in a finite domain the Dirichlet problem and the mixed problem have unique solutions. The Neumann problem with the condition (2.4.2) has solution under the condition

$$\int_{\partial G} g(x, y) \, ds = 0 \,,$$

and such a solution is determined to within an additive constant.

For infinite plane domains the boundary-value problems for harmonic functions are formulated differently for different types of domains. For example, for an infinite domain, which is the exterior of a finite contour ∂G the theorems of existence of the solutions of boundary-value problems state:

1. The unique solutions limited at the point at infinity exist for the Dirichlet problem and mixed boundary-value problem;

2. The limited solution of the Neumann problem with the boundary condition (2.4.2) exists if

$$\int_{\partial G} g(x, y) \, ds = 0 \, .$$

This solution is accurately determined except for the additive constant. If

$$\int_{\partial G} g(x, y) \, ds = C \neq 0 \, ,$$

the external Neumann problem has unlimited solution which grows as $C/(2\pi) \ln R$ with $|z| = R \to \infty$.

The effective method of solution of boundary-value problems for harmonic functions in the plane domains of the complex shape is the method of conformal mapping. This method consists in the conformal mapping of the considered complex domain onto some simple domain (a circle, a circle ring, a half-plane or a strip), for which the solution of the corresponding problem is known. Due to the invariance of a harmonic function under the conformal mapping (see Section 1.8) it is the solution of the boundary-value problem in the initial complex domain.

2.5   The Construction of a Green Function of the Dirichlet Problem

Let G be a singly connected domain of the u, v plane with the boundary L, $M_0(u_0, v_0)$ a fixed point of the domain. Designate an arbitrary point of the domain as $M(u, v)$ and assign a complex number $w = u + iv$ to the point M, and a number $w_0 = u_0 + iv_0$ to point M_0.

The *Green function of the Dirichlet problem* is the function determined by conditions

$$\nabla^2 \mathcal{G}(M, M_0) = 0 \, , \quad M \neq M_0 \, ,$$

$$\mathcal{G}|_L = 0 \, , \tag{2.5.1}$$

$$\mathcal{G}(M, M_0) = -\frac{1}{2\pi} \ln R + \Phi(u, v) \, ,$$

where $R = |MM_0|$ is the distance between the points M and M_0; $\Phi(u, v)$ — some function harmonic in the domain G. In the case when the domain G is unlimited, the natural condition of limitation of the function $\mathcal{G}(M, M_0)$ at the point at infinity is applied. There exists the unique function $\mathcal{G}(M, M_0)$ determined by the conditions (2.5.1).

The function $\mathcal{G}(M, M_0)$ expresses:

1. The potential of the constant electric current inside a flat plate in the case when the boundary of the plate Γ is grounded and a point electrode is connected at the point M_0;

2. The potential of an electrostatic field of an infinite charged fiber in the presence of an infinite grounded cylinder;

3. The stationary distribution of the temperature in a plane plate when the contour of the plate is sustained at zero temperature and the point source of heat is placed at the point M_0;

4. Vector potential $V(u, v)$ of flow velocity of ideal liquid caused by a vortex fiber in the presence of an impenetrable cylinder with directrix line L.*

Let the function

$$z = F(w, w_0) \qquad (2.5.2)$$

conformally map the domain G onto the unit circle $|z| < 1$ so that the point w_0 would be mapped into the center of the circle. It is not difficult to check that the function

$$\mathcal{G}(M, M_0) = -\frac{1}{2\pi} \ln |F(w, w_0)|$$

satisfies all the conditions (2.5.1). The function (2.5.2) maps the net of level lines of the function $\mathcal{G}$ and its orthogonal trajectories† onto the polar net of lines in the circle $|z| < 1$. The inverse function $w = f(z)$ conformally maps the circle $|z| < 1$ onto the domain G, the polar net in the circle is mapped into the relief map of the Green function for the Dirichlet problem.

To visualize the Green function it is necessary to find the function which conformally maps the circle $|z| < 1$ onto the domain G so that the center of the circle would be mapped into the point w_0 and to map the polar net by the function.

Let some function $w = f(t)$ having mapped the circle $|t| < 1$ onto the domain G be known. Designate as t_0 the preimage of the point w_0 in this mapping: $t_0 = f^{-1}(w_0)$.

The mapping of the circle $|z| < 1$ onto the circle $|t| < 1$ under which the point $z = 0$ is mapped into the point t_0 is realized by the linear-fractional function $t = (z - t_0)/(z\overline{t_0} - 1)$. Thus, the desired mapping of the circle $|z| < 1$ onto the domain G, when the point $z = 0$ is mapped into the point

---

*On the impenetrable wall the scalar potential of the velocity field $U(u, v)$ satisfies the condition $v = \partial U/\partial n|_L = 0$ whence due the Cauchy-Riemann conditions $\partial V/\partial \tau|_L = 0$, where $\partial/\partial \tau$ is the derivative along the tangent to the curve L. The last condition yields $V|_L = \text{const}$, and an arbitrary constant can be assumed zero without loss of generality.

†An orthogonal trajectory of a family of the smooth non-intersecting plane curves is the line which intersects each line of the family at the right angle.

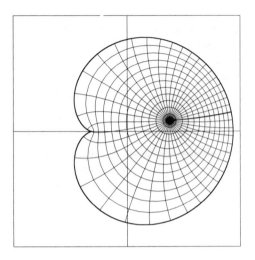

FIGURE 2.2

w_0, is realized by the composition of functions

$$t = \frac{z - t_0}{\overline{zt_0} - 1}, \quad w = f(t), \tag{2.5.3}$$

where $|t_0| < 1$. Taking an expression of the function from the catalog 1 of
the program CONFORM it is possible to construct the relief map of the
Green function for any domain of the catalog. When using the program
CONFORM the position of the point w_0 is not given, but it is defined by
fitting of the complex parameter t_0.

Example 2.1
The function

$$t = \frac{z - t_0}{zt_0 - 1}, \quad w = 2t + t^2$$

maps the polar net in the circle $|z| < 1$ onto the relief map of the Green
function for the domain bounded by the cardioid. The mapping is presented
in Fig. 2.2 with $t_0 = Re^{ia}$, $R = 0.5$, $a = 0.1$. The position of the source
can be varied changing the complex parameter t_0 ($|t_0| < 1$).

    If the domain G is an exterior of the finite contour and the function
$w = f(t)$ which maps the exterior of the circle $|t| > 1$ onto the domain G is
known, the relief map of the Green function in the domain G is built as the
mapping of the polar net in the circle $|z| < 1$, realized by the composition of
the functions (2.5.3) where $|t_0| > 1$ should be taken. A set of such domains
is presented in Catalog 2 of CONFORM program.

If the domain G is a curvilinear angular domain and the function $w = f(t)$ which maps the half-plane $\operatorname{Im} t > 0$ onto the domain G is known, the relief map of the Green function for the domain is the mapping of the polar net in the circle $|z| < 1$ carried out by the composition of functions

$$t = \frac{z\overline{t_0} - t_0}{z - 1}, \quad w = f(t), \tag{2.5.4}$$

where $\operatorname{Im} t_0 > 0$. With the help of formulae (2.5.4) the Green function for any domain from catalog 3 can be visualized.

2.6 The Green Function of the Neumann Problem

Let an infinite domain G be an exterior of the finite contour L, M_0 be an interior point of the domain. *The Green function of the Neumann problem in the domain G is the function* $\mathcal{H}(M, M_0)$, which satisfies the conditions:

$$\nabla^2 \mathcal{H} = 0, \quad M \neq M_0,$$

$$\left. \frac{\partial \mathcal{H}}{\partial n} \right|_L = 0, \tag{2.6.1}$$

$$\mathcal{H} = -\frac{1}{2\pi} \ln R + \Phi(u, v),$$

where Φ designates some function harmonic in the domain G and limited at the point at infinity. The conditions (2.6.1) determine the function $\mathcal{H}$ to within an additive constant.

The function $\mathcal{H}(M, M_0)$ expresses, for example:

1. The potential of electric current in the infinite plate (the domain G) if the boundary of the plate is electrically isolated and a point electrode is placed at the point M_0;

2. Stationary distribution of the temperature in the infinite homogeneous plate if the boundary of the plate L is heat isolated and the point source of the heat is placed at the point M_0;

3. The vector potential of magnetic field of linear current in the presence of infinite cylinder made of magnetic material with infinitely large magnetic permittivity.

The Green function of the Neumann problem can be constructed with the help of conformal mapping. Let the function $z = F(w, w_0)$ map the domain G onto the exterior of some segment of the real axis $(p - 1, p + 1)$ (where $p < -1$) with the points $w = w_0$ and $w = \infty$ being mapped into the points $z = 0$ and $z = \infty$, respectively. The function

$$\mathcal{H}(M, M_0) = -\frac{1}{2\pi} \ln |F(w, w_0)| + C$$

satisfies all the conditions (2.6.1). The function $F(w, w_0)$ maps the relief map of the function $\mathcal{H}(M, M_0)$ onto the polar net.

To construct the function $F(w, w_0)$ the following algorithm is proposed. Let the domain G be mapped onto the exterior of the unit circle by some function $t = \Phi(w)$. Let the point w_0 be mapped into the point $t_0 = \Phi(w_0)$. The exterior of the circle is mapped onto that of the segment of the real axis by the sequence of transformations

$$s = te^{-ia},$$

$$r = \frac{1}{2}\left(s + \frac{1}{s}\right),$$

$$z = r - r_0,$$

where $t_0 = Re^{ia}$, $r_0 = 1/2(R + 1/R)$.

The inverse mapping of the exterior of the segment onto the domain G with the needed correspondence of the points is carried out by the sequence of transformations

$$r = z + r_0,$$

$$s = r + \sqrt{r-1}\,\sqrt{r+1},$$

$$t = se^{ia}, \tag{2.6.2}$$

$$w = f(t),$$

where $f(t)$ is the function inverse to the function $t = \Phi(w)$. With the help of the program CONFORM it is possible to construct by the formulae (2.6.2) the relief map of the function $\mathcal{H}(M, M_0)$ as the mapping of polar net. Any function from Catalog 2 can be taken as $f(t)$.

Example 2.2

The relief map of the Green function of the Neumann problem in the exterior of an ellipse is shown in Fig. 2.3. The figure is created by formulae (2.6.2) with $\zeta = 1.7t$, $w = 1/2(\zeta + 1/\zeta)$, $r_0 = 1.5$, $a = 0.7$.

The Green function of the Neumann problem in a curvilinear angular domain can be constructed with the help of a conformal mapping of the domain onto the plane z with the cut along the ray of the real axis. Let

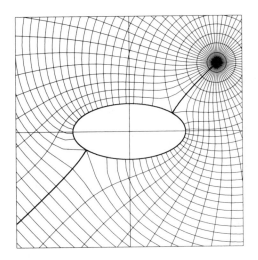

FIGURE 2.3

the function $z = F(w, w_0)$ conformally map the curvilinear angular domain G onto the plane z with the cut along the ray $(-\infty, -1)$, the points $w = w_0$ and $w = \infty$ being mapped into the points $z = 0$ and $z = \infty$, correspondingly. Then

$$\mathcal{H}(M, M_0) = -\frac{1}{2\pi} \ln |F(w, w_0)| + C,$$

where C is an arbitrary constant.

The function $F(w, w_0)$ maps the relief map of the Green function onto the polar net of the plane z. It can be constructed as follows.

Let the function $t = \Phi(w)$ map the curvilinear angular domain G onto the half-plane $\operatorname{Im} t > 0$ so that the point $w = \infty$ is mapped into the point $t = \infty$. Designate $t_0 = \Phi(w_0) = a + ib$. The sought mapping of the half-plane $\operatorname{Im} t > 0$ onto the plane z with the cut along the ray $(-\infty, -1)$ is realized by the function

$$z = -1 - \frac{(t-a)^2}{b^2}.$$

The inverse mapping of the plane z with the cut onto the curvilinear angular domain G is realized by the composition of functions

$$t = a + ib\sqrt{z+1},$$

$$w = f(t), \tag{2.6.3}$$

where $f(t)$ is the function inverse to the function $t = \Phi(w)$.

The function (2.6.3) maps the polar net of lines in the plane z into the relief map of the Green function. This function allows us to visualize the

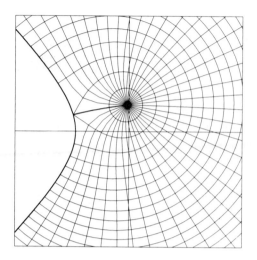

FIGURE 2.4

Green function of the Neumann problem for any curvilinear angular domain from Catalog 3. The position of the source M_0 can be changed by fitting of real parameters a and b.

Example 2.3

The relief map of the Green function of the Neumann problem in the domain bounded by one of the branches of hyperbola is presented in Fig. 2.4. The image is created with the help of the function

$$t = a + ib\sqrt{z+1}\,,$$

$$r = -it\,,$$

$$s = \arccos r\,,$$

$$w = \cos(ps)$$

with $a = 0.6$, $b = 0.1$, $p = 1.5$.

2.7   The Green Function of a Mixed Boundary-Value Problem

Let the boundary of a singly connected domain G consist of two connected parts L_1 and L_2. The *Green function of the mixed boundary-value problem* in the domain G is the function $\mathcal{S}(M, M_0)$, which satisfies the conditions:

$$\nabla^2 \mathcal{S} = 0, \quad M \neq M_0,$$

$$\mathcal{S}|_{L_1} = 0,$$

$$\frac{\partial \mathcal{S}}{\partial n}\bigg|_{L_2} = 0, \qquad (2.7.1)$$

$$\mathcal{S}(M, M_0) = -\frac{1}{2\pi} \ln R + \Phi(u, v),$$

If the domain G is unlimited, the condition of limitation of the function $\mathcal{S}(M, M_0)$ at the point at infinity is added to conditions (2.7.1). The function $\mathcal{S}(M, M_0)$ is uniquely determined by the mentioned conditions.

The function $\mathcal{S}(M, M_0)$ expresses the potential of electric current in the homogeneous conductive plate if the part of its boundary L_1 is grounded, the other part L_2 is electrically isolated, and a point electrode is placed at the point M_0.

The function $\mathcal{S}(M, M_0)$ can be constructed with the help of the conformal mapping of the domain G onto a circle with a cut along a segment of the radius. Let the function $z = F(w, w_0)$ carry out the conformal mapping of the domain G onto the circle $|z| < 1$ with the cut along the segment $(-1, -a)$, so that the point w_0 is mapped into the center of the circle, the curve L_1 is mapped onto the circumference $|z| = 1$, and the curve L_2 — onto the cut $(-1, -a)$. The function

$$\mathcal{S}(M, M_0) = -\frac{1}{2\pi} \ln |F(w, w_0)|$$

satisfies all the conditions (2.7.1).

The equivalent of the problem (2.7.1) is the problem on the distribution of the electric current in the round plate with grounded edge and electrically isolated cut along the segment of the radius. This current is created by the point electrode in the center of the circle. It is obvious that the polar net is the net of equipotentials and streamlines in this plate. Thus, the function $z = F(w, w_0)$ maps the relief map of the Green function onto the polar net. The inverse function $w = F^{-1} = f(z)$ maps the polar net in the circle onto the relief map of the function $\mathcal{S}(M, M_0)$ in the domain G.

The conformal mapping of the domain G onto the circle with the cut is constructed in the following way. The domain G can be conformally mapped by some function $t = \Phi(w)$ onto the half-plane $\text{Im}\, t > 0$ that the curve L_1 would be mapped into the negative real half-axis and the curve L_2 — into the positive one.*

---

*Indeed, due to the Riemann theorem the domain G can be conformally mapped by some function $\xi = \varphi(w)$ onto the half-plane $\text{Im}\, \xi > 0$. If the curve L_1 is mapped into a finite segment of the real axis (a, b), the sought for mapping onto the half-plane $\text{Im}\, t > 0$

Let the image of the point w_0 be the point $t_0 = \Phi(w_0) = Re^{ia}$. The circle $|z| < 1$ with a cut along some segment of the radius $(-1, -a)$ can be conformally mapped onto the half-plane $\mathrm{Im}\, t > 0$ so that the cut would be mapped onto the positive real half-axis. This mapping is carried out by the sequence of transformations:

$$r = \frac{1+z}{1-z}, \quad s = i\sqrt{r^2 - b^2}, \quad t = A\frac{b+s}{b-s},$$

where $b = (1-a)/(1+a)$, $A > 0$.

Arbitrary parameters A and a (or b) can be fitted so that the center of the circle would be mapped into the point t_0. For the purpose there ought to be taken $A = R = |t_0|$, $b = \cos\alpha/2 = \cos(\arg t_0/2)$. Therefore, the mapping of the circle $|z| < 1$ with the cut along the segment of the radius onto the domain G is carried out by composition of functions

$$r = \frac{1+z}{1-z},$$

$$s = i\sqrt{r^2 - b^2},$$

$$t = R\frac{b+s}{b-s}, \tag{2.7.2}$$

$$w = f(t),$$

where $f(t)$ is the function inverse to the function $t = \Phi(w)$.

The formulae (2.7.2) allow us to construct the relief map of the Green function of the mixed boundary-value problem for any curvilinear angular domain from Catalog 3. For singly connected domains of other types it is necessary at first to map an adequate canonical domain onto a half-plane and then to use the formula (2.7.2).

Example 2.4

The relief map of the Green function of the mixed boundary-value problem in an angular domain is shown in Fig. 2.5. The map represents the net of streamlines and equipotentials for the distribution of electric current caused

---

will be

$$t = \frac{\xi - a}{-\xi + b}.$$

If the curve L_2 is mapped into a finite segment (a, b), the necessary mapping of the half-plane $\mathrm{Im}\,\xi > 0$ onto the half-plane $\mathrm{Im}\, t > 0$ is carried out by the function

$$t = \frac{\xi - b}{\xi - a}.$$

At last, if the curve L_1 is mapped into a ray of the real axis (a, ∞), the desired linear-fractional transformation will be $t = 1/(a - \xi)$ and if the image of the curve L_1 is the ray $(-\infty, b)$, the sought for transformation will be the linear one: $t = \xi - b$.

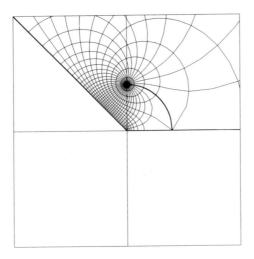

FIGURE 2.5

by a point electrode in the infinite conductive plate. This plate has angular shape with apex angle πp. One side of the plate is electrically isolated and the other side has zero potential. The relief map represents the mapping of the polar net of lines in the circle $|z| < 1$ carried out by the composition of functions (2.7.2) with $f(t) = t^p$, $b = 0.818$, $R = 1$, $p = 0.75$.

2.8 Point Source and Sink in the Dirichlet Problem

Let the boundary L of a conductive plate G have zero potential and let point electrodes of opposite polarity (the "source" and the "sink") be placed at the points $M_0(u_0, v_0)$ and $M_1(u_1, v_1)$. The potential of electric current created by these electrodes satisfies the conditions

$$\nabla^2 U = 0, \quad M \neq M_0, \quad M \neq M_1,$$

$$U|_L = 0, \tag{2.8.1}$$

$$U = -\frac{1}{2\pi} \ln R_0 + \frac{1}{2\pi} \ln R_1 + \Phi(u, v),$$

where $R_0 = |MM_0|$, $R_1 = |MM_1|$, $\Phi(u, v)$ is a harmonic function in the domain G. The solution of the problem (2.8.1) in the singly connected domain G can be built with the help of conformal mappings.

Designate $w_0 = u_0 + iv_0$, $w_1 = u_1 + iv_1$, $w = u + iv$. Let the function $z = F(w, w_0, w_1)$ conformally map the domain G onto the exterior of an arc of the circumference $|z| = 1$, $\pi/2 - \beta < \arg z < \pi/2 + \beta$ so that the point w_0 would be mapped into the point $z = 0$ and the point w_1 — into the point at infinity. The function

$$U = -\frac{1}{2\pi}\ln|F(w, w_0, w_1)|$$

satisfies all the conditions of the problem (2.8.1). The function $z = F(w, w_0, w_1)$ can be built as follows. Let the function $\xi = \Phi(w)$ map the domain D onto the exterior of the unit circle $|\xi| > 1$ and let the points $\xi_0 = \Phi(w_0)$ and $\xi_1 = \Phi(w_1)$ be images of the points w_0 and w_1. The exterior of the circle $|\xi| > 1$ can be mapped onto the exterior of the circle $|r| > 1$ so that the point ξ_1 would be mapped into the point $r = \infty$:

$$r = \frac{\overline{\xi_1}\xi - 1}{-\xi + \xi_1}.$$

Let the image of the point ξ_0 be the point $r_0 = Re^{i\gamma}$. Designate $R = \sec\alpha$, i.e., introduce the parameter $\alpha = \arccos(1/R)$. A sequence of the linear transformations

$$s = -ire^{-i\gamma}, \quad t = \frac{s + i\cos\alpha}{\sin\alpha}$$

maps the considered domain onto the exterior of a circumference $|t - i\cot\alpha| > 1/\sin\alpha$ passing through the points $t = \pm 1$. At that the point r_0 (the source) is mapped into the point $t_0 = -i\tan\alpha$.

The Zhukovskii function

$$\zeta = \frac{1}{2}\left(t + \frac{1}{t}\right)$$

in accordance with the Subsection 1.7.4 maps the exterior of the circle onto the exterior of the circumference arc with the ends at points $\zeta = \pm 1$. The point $t_0 = -i\tan\alpha$ is mapped into the center of the circumference — the point $\zeta_0 = i\cot 2\alpha$. Finally the linear transformation

$$z = (\zeta - i\cot 2\alpha)\sin 2\alpha$$

maps the considered domain onto the exterior of the arc of the unit circumference $|z| = 1$, $\pi/2 - \beta < \arg z < \pi/2 + \beta$ (where $\beta = \pi - 2\alpha$), the point $\zeta_0 = i\cot 2\alpha$ being mapped into the origin.

The inverse mapping of the exterior of the arc of the unit circumference onto the domain G is realized by the composition of functions

$$\zeta = \frac{z + i\cos 2\alpha}{\sin 2\alpha},$$

$$\tau = e^{i\beta/2}\sqrt{\frac{\zeta - 1}{\zeta + 1}}\,e^{i\beta},$$

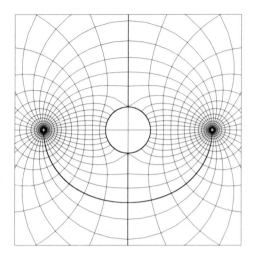

FIGURE 2.6

$$t = \frac{1+\tau}{1-\tau},$$

$$s = t\sin\alpha - i\cos\alpha,$$

$$\xi = \frac{r\xi_1 + 1}{r + \overline{\xi_1}},$$

$$w = f(\xi),$$

(2.8.2)

where $f(\xi) = \Phi^{-1}$ is the function inverse to the function $\xi = \Phi(w)$.

The composition of functions (2.8.2) maps the polar net of lines in the z plane onto the net of equipotentials and streamlines of the considered problem. With the help of the function (2.8.2) it is possible to visualize the solution of the problem (2.8.1) for any domain presented in the Catalog 2 of the program CONFORM. For the domains of other types one should first map the domain onto the exterior of a circle and then use the considered algorithm of the mapping of the exterior of a circle onto the exterior of the circumference arc.

Example 2.5

The map of equipotentials and streamlines for the distribution of the electric current $\mathbf{J}$ created by two point electrodes in the exterior of the circumference with zero potential is presented in Fig. 2.6. The mapping is constructed by the formulae (2.8.2) with $\alpha = \pi/6$, $\gamma = 0$, $\xi_1 = -\cot(\beta/4)$, $w(\xi) = \xi$. To obtain an even net the variable z was considered as intermediate: $z = e^{z_1}$, where the variable z_1 belongs to the strip $-\infty < \mathrm{Re}\, z_1 < \infty$, $-\pi < \mathrm{Im}\, z_1 < \pi$.

Simultaneously the Fig. 2.6 visualizes the harmonic conjugate vector field $\mathbf{H} = \mathbf{J}^*$. The field $\mathbf{H}$ can be interpreted as a magnetic field of two parallel wires with oppositely directed electric currents in the presence of a superconductive circular cylinder.

2.9 Point Source and Sink in the Neumann Problem

Let a singly connected domain G be the uniform electroconductive plate with electrically isolated boundary L and let two point electrodes with opposite polarities be placed at the points $M_0(u_0, v_0)$ and $M_1(u_1, v_1)$. The potential of electric current $U(M, M_0, M_1)$ in the domain G satisfies the conditions

$$\nabla^2 U = 0, \quad M \neq M_0, \quad M \neq M_1,$$

$$\left. \frac{\partial U}{\partial n} \right|_L = 0, \tag{2.9.1}$$

$$U = -\frac{1}{2\pi} \ln R_0 + \frac{1}{2\pi} \ln R_1 + \Phi(u, v)$$

(see previous section for designations). The function U is determined by the conditions (2.9.1) to within a constant term.

The function $U(M, M_0, M_1)$ can be built with the help of conformal mappings. Let the function $z = F(w, w_0, w_1)$ conformally map the domain G onto the exterior of a segment of the real axis $-p - 1 < \operatorname{Re} z < -p + 1$ (where $p > 1$), the point w_0 being mapped into the point $z = 0$ and the point w_1 — into the point $z = \infty$. The function

$$U(M, M_0, M_1) = -\frac{1}{2\pi} \ln |F(w, w_o, w_1)| + C$$

satisfies all the conditions (2.9.1) (here C is an arbitrary constant).

The function $F(w, w_0, w_1)$ is constructed as follows. Let the function $\xi = \Phi(w)$ map the domain G onto the exterior of the unit circle and the points w_0 and w_1 — into the points ξ_0 and ξ_1 correspondingly.

The linear-fractional function

$$r = \frac{\xi \overline{\xi_1} - 1}{-\xi + \xi_1}$$

maps the considered domain onto the exterior of the circle $|z| > 1$, the point $\xi = \xi_1$ being mapped into the point $r = \infty$. Let the point $r_0 = Re^{i\gamma}$ be the image of the source ξ_0. After the turn $s = re^{-i\gamma}$ (where $\gamma = \arg r_0$) the considered domain is mapped onto the exterior of the circle $|z| > 1$, the source (the point r_0) being mapped into the point $s_0 = R$ ($R > 1$). The Zhukovskii function

$$\zeta = \frac{1}{2}\left(s + \frac{1}{s}\right)$$

maps the considered domain onto the exterior of the segment $[-1, 1]$, the point $s_0 = R$ being mapped into the point on the real axis $p = 1/2(R+1/R)$. The sought for mapping of the domain G onto the plane with the cut along a segment of the real axis is realized by the function

$$z = \zeta - p.$$

The inverse mapping of the plane z with the cut onto the domain G is carried out by the sequence of transformations

$$\begin{aligned}
\zeta &= z + p, \\
s &= \zeta + \sqrt{\zeta^2 - 1}, \\
r &= s\,e^{i\gamma}, \\
\xi &= \frac{r\xi_1 + 1}{r + \overline{\xi_1}}, \\
w &= f(\xi),
\end{aligned}$$

(2.9.2)

where $w = f(\xi) = \Phi^{-1}$ is the function inverse to $\xi = \Phi(w)$. The function (2.9.2) maps the polar net of the z plane onto the net of equipotentials and streamlines for the electric current $\mathbf{J}$ created by two point sources in the domain G. The position of sources can be changed by varying the parameters p, γ and ξ_1.

With the help of the function (2.9.2) it is possible to visualize the function $U(M, M_0, M_1)$ for any domain presented in the Catalog 2 of the program CONFORM. For the domains of other types it is necessary first to map the domain onto the exterior of the circle and then to use the considered algorithm of the mapping of an exterior of the circle onto that of the segment.

Example 2.6

The net of equipotentials and streamlines for the electric current $\mathbf{J}$ created by two point electrodes in the exterior of the electrically isolated circle is presented in Fig. 2.7. The mapping is constructed by the formulae (2.9.2) with $p = 1.1$, $\gamma = 0$, $\xi_1 = -(p + \sqrt{p^2 - 1})$, $w(\xi) = \xi$. To obtain an even net the variable z was used as an intermediate one: $z = e^{z_1}$, where z_1 belongs to the strip $-\infty < \mathrm{Re}\, z_1 < \infty$, $-\pi < \mathrm{Im}\, z_1 < \pi$.

Simultaneously the Fig. 2.6 visualizes the harmonic conjugate vector field $\mathbf{H} = \mathbf{J}^*$. The field $\mathbf{H}$ can be interpreted as a magnetic field of two parallel straight wires with oppositely directed electric currents in the presence of

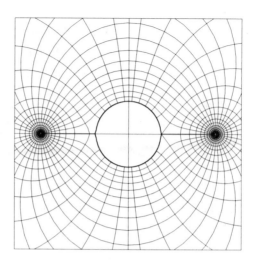

FIGURE 2.7

an infinite round cylinder made of the magnetic material with infinitely large magnetic permittivity.

2.10   Plane Robin Problem

The *Robin problem* is the problem on the distribution of the electric charge on the surface of a charged conductive body. The plane Robin problem is the problem on the electric field of infinite charged cylinder having an arbitrary cross-section.

Let a plane singly connected domain G be an exterior of the finite contour L. The plane Robin problem for the cylinder with the directrix L is formulated as a boundary-value problem for the potential of the electric field[*]

$$\nabla^2 U = 0 \,,$$

$$U|_L = 0 \,,$$

$$\int_L \frac{\partial U}{\partial n}\, ds = 1 \,. \qquad (2.10.1)$$

The problem (2.10.1) has a unique solution. This solution grows as $1/(2\pi)\ln R$ with $R \to \infty$.

---

[*]The surface of a conductive body is an equipotential: $U|_L = $ const. This constant can be assumed as zero. The latest condition (2.10.1) means that the unit length of the cylinder has the unit charge.

Let the function $z = F(w)$ map the domain G conformally onto the exterior of the circle $|z| > 1$, the point $w = \infty$ being mapped into the point $z = \infty$. The function

$$U(M) = \frac{1}{2\pi} \ln |F(w)|$$

satisfies all the conditions (2.10.1). The inverse function $w = f(z)$ (where $f = F^{-1}$) maps the polar net of lines in the exterior of the circle $|z| > 1$ onto the relief map of the function $U(M)$. In such a way the solution of the Robin problem for any domain presented in the Catalog 2 of the program CONFORM can be visualized. Examples of relief maps for the solution of the plane Robin problem in various domains was presented earlier in Fig. 1.43, 1.44, 1.48, 1.51, 1.52, 1.121, and 1.133. Each figure visualizes the electric field $\mathbf{E}$ of a charged cylinder. Simultaneously they can be interpreted as a net of equipotentials and force lines of harmonic conjugate field $\mathbf{H} = \mathbf{E}^*$ which is a magnetic field created by the electric current flowing along the superconductive cylinder.

2.11   The Flow in a Curvilinear Angular Domain. Streaming Around an Infinite Curve

Let a domain G be curvilinear angular domain of the w plane bounded by an infinite curve L. Let the apex angle of the domain G (the angle at the point at infinity — see Subsection 1.1.4) equal πp. The domain G can be conformally mapped with the help of some function $z = F(w)$ onto the half-plane Im $z > 0$, the point $w = \infty$ being mapped* into the point $z = \infty$.

The function $z = F(w)$ can be considered as a complex potential of the velocity field of a liquid in the half-plane Im $z > 0$. The streamlines and equipotentials of the flow is a Cartesian net. The inverse function $w = f(z)$ maps the half-plane onto the domain G, the Cartesian net in the half-plane being mapped onto the net of equipotentials and streamlines for the velocity field $\mathbf{v}$ of the liquid streaming around the curve L. Thus, the vector field $\mathbf{v}$ can be visualized for any domain presented in the Catalog 3 of the program CONFORM. Examples of visualization of flows in curvilinear angular domains are presented above in Fig. 1.41, 1.42, 1.59, 1.60, etc.

Simultaneously nets of lines depicted in these figures visualize harmonic conjugate vector field $\mathbf{A} = \mathbf{v}^*$. The field $\mathbf{A}$ can be interpreted as an electric field of an infinite electrode, which is a cylinder with directrix L (another electrode is at the point at infinity). Another interpretation is as follows:

---

*Such a mapping exists but is not unique. Indeed, the function $z = AF(w) + B$ (where A, B are the real constants, $A > 0$) carries out the same mapping.

A is the magnetic field near a pole of an infinite magnet made of ideal magnetic material (other pole is at infinity).

The function $z = F(w)$ which maps the curvilinear angular domain with the apex angle πp onto the half-plane has the asymptotic

$$F(w) = O(w^{1/p})$$

as $|w| \to \infty$.

The modulus of the liquid velocity $|\mathbf{v}|$ has the asymptotic

$$|\mathbf{v}| = |F'(w)| = O(w^{1/p-1}).$$

Therefore, character of the motion of the liquid at infinity depends on the value of apex angle of the curvilinear angular domain: if $0 < p < 1$ then $|\mathbf{v}| \to \infty$ with $|w| \to \infty$; if $1 < p < 2$ then $|\mathbf{v}| \to 0$ with $|w| \to \infty$. Only when $p = 1$ (i.e., in the case of curvilinear half-plane) the liquid has the finite velocity at the point at infinity. In this case the complex potential of the velocity field $\mathbf{v}$ in the curvilinear half-plane has the form

$$z = A\,F(w)\,, \tag{2.11.1}$$

where $A = v_0/|F'(\infty)|$, $F'(\infty) = \lim\limits_{w \to \infty} F'(w)$, v_0 is an absolute value of the liquid velocity at the point at infinity.

For curvilinear angular domains with the apex angle differing from π the normalization condition (2.12.1) is not valid. In this case for the unique determination of the velocity field of the liquid it is necessary to set an absolute value of the flow velocity v_0 in some finite point of the domain w_0. The complex potential of this vector field is

$$z = A\,F(w)\,,$$

where $A = v_0/|F'(w_0)|$.

2.12   The Flow in a Curvilinear Strip

Let a domain G in the plane w be a curvilinear strip bounded by two infinite curves L_1 and L_2. Designate the boundary points of this strip lying at infinity as A_1 and A_2.

Consider the problem on the potential flow of the ideal liquid in the channel of the constant depth with the walls L_1 and L_2. It can be formulated as the boundary-value problem for the stream function $V(u, v)$:

$$\nabla^2 V = 0\,,$$

$$V|_{L_1} = 0\,, \tag{2.12.1}$$

$$V|_{L_2} = N,$$

where N is the flow rate in the channel. This problem is solved by the conformal mapping of the domain G onto the adequate canonical domain — the straight line strip.

The domain G can be conformally mapped with the help of the function $z = F(w)$ onto the strip $0 < \text{Im } z < \pi$, the points A_1 and A_2 being mapped* into the points $z = \pm\infty$.

The velocity field of the ideal liquid in the strip $0 < \text{Im } z < \pi$ has the complex potential $N/\pi z$. The Cartesian net is the net of its equipotentials and streamlines.

Assuming $N = \pi$ we obtain the complex potential of the flow in the curvilinear strip: $z = F(w)$. The inverse function $w = F(z)$ conformally maps the strip $0 < \text{Im } z < \pi$ onto the domain G and the Cartesian net — onto the net of streamlines and equipotentials for the velocity field $\mathbf{v}$ of the ideal liquid in the domain G. Thus the velocity field $\mathbf{v}$ can be visualized for any curvilinear strip in the Catalog 4 of the program CONFORM. Examples of such mappings are presented above in Fig. 1.58, 1.61, 1.62, 1.63, etc. Each mapping in these figures visualizes the harmonic conjugate field $\mathbf{A} = \mathbf{v}^*$ together with the field $\mathbf{v}$. The field $\mathbf{A}$ can be interpreted as:

(a) a stationary flux of heat $\mathbf{q}$ in the curvilinear wall of homogeneous material, if the surface L_1 has zero temperature and the surface L_2 has temperature $T_0 = \text{const}$;

(b) magnetic field in the gap of electromagnet between the poles L_1 and L_2 made of the magnetic material with infinitely large magnetic permittivity;

(c) electric field in the curvilinear capacitor with the plates L_1 and L_2 if one plate has zero potential and the other — the potential $U_0 = \text{const}$.

2.13 The Flow in a Curvilinear Strip with N Branches

Let a domain G in the plane w be a curvilinear strip with N branches, bounded by infinite curves $L_1, L_2, \dots, L_N$ without common points except for the point at infinity. Designate the boundary points of the domain G lying at infinity as $A_1, A_2, \dots, A_N$ (the points are numbered in the positive sense of boundary traversal; the point A_k is the end of L_{k-1} curve and the beginning of L_k curve). Call the points A_k the branches of the strip G.

---

*Such a mapping exists but is not unique. Indeed, the function $z_1 = F(w) + C$ (where C is the real constant) carries out the same mapping.

Consider two branches A_k and A_m having a source and a sink with intensity N respectively. A flow of the ideal liquid in the domain G created by such source-sink pair is called the flow without branching. It can be visualized by the conformal mapping of the domain G onto the straight line strip.

The domain G can be conformally mapped onto the strip $0 < \operatorname{Im} z < \pi$ so that the points A_k and A_m would be mapped into the points $z = \pm\infty$. First it is necessary to map the domain G with the help of the function $\xi = F(w)$ onto the half-plane $\operatorname{Im} \xi > 0$. Let the points A_k and A_m be mapped into the points of the real axis $\xi = a_k$ and $\xi = a_m$. If one of these points, say a_k, is the point at infinity then the mapping of the half-plane onto the strip $0 < \operatorname{Im} z < \pi$ with the necessary correspondence of the points is realized by the composition of functions

$$t = \xi - a_m, \quad z = \ln t.$$

If both of the points a_k and a_m are finite and $a_m > a_k$ then the mapping is realized by the composition of functions

$$t = \frac{\xi - a_m}{\xi - a_k}, \quad z = \ln t.$$

The complex potential of the flow in the strip $0 < \operatorname{Im} z < \pi$ is equal to $N/\pi\, z$. Assuming $N = \pi$ consider the variable z as the complex potential of the flow without branching in the domain G. The inverse mapping of the strip $0 < \operatorname{Im} z < \pi$ onto the domain G is realized by the composition of functions

$$t = e^z, \tag{2.13.1}$$

$$\xi = \frac{ta_k - a_m}{t - 1}, \tag{2.13.2}$$

$$w = f(\xi), \tag{2.13.3}$$

where $f(\xi) = F^{-1}$ is the function inverse to the function $\xi = F(w)$. The formulae (2.13.1)–(2.13.3) are written for the case when both of the points a_k and a_m are finite. If $a_k = \infty$ then instead of linear-fractional function (2.13.2) one should write

$$\xi = t + a_m.$$

The function (2.13.1)–(2.13.3) maps the Cartesian net in the strip $0 < \operatorname{Im} z < \pi$ onto the net of equipotentials and streamlines for the flow without branching in the domain G. With the help of it one can visualize the velocity field $\mathbf{v}$ of the ideal liquid in any domain included into the Catalog 5 of the program CONFORM (mode **Mapping**). For each domain in the Catalog 5 the numbers a_k are indicated, that are preimages of strip branches.

Examples of visualization of flows without branching in curvilinear strips with several branches are presented above in Fig. 1.82 and 1.83. In these figures the harmonic conjugate field $\mathbf{A} = \mathbf{v}^*$ is visualized together with the velocity field of ideal liquid $\mathbf{v}$. The field $\mathbf{A}$ can be interpreted as the electric field in a capacitor with one plate formed by the curves L_k, $L_{k+1}, \ldots, L_{m-1}$ and the other plate — by the rest $N - m + k$ curves.

2.14 The Flow Running onto an Infinite Curve and Branching on It

Let L be an infinite curve of the plane w bounding a curvilinear angular domain G with the apex angle 2π; w_0 — some point of the curve L. Consider a flow of the ideal liquid running onto the curve L and branching at the point w_0 into two flows streaming around the different sides of the curve L. The problem of visualization of the flow with branching can be solved with the help of the conformal mapping of the domain G onto the plane z with a cut along the negative real half-axis, with the points $w = \infty$ and $w = w_0$ being mapped into the points $z = \infty$ and $z = 0$ correspondingly.

This mapping can be built as follows. Let the function $\xi = F(w)$ map the domain G onto the half-plane Im $\xi > 0$ so that the point $w = \infty$ is mapped into the point $\xi = \infty$. Designate the image of the point w_0 as p: $p = F(w_0)$. The mapping of the half-plane Im $\xi > 0$ onto the plane z with the cut along the negative half-axis with the necessary correspondence of the points is realized by the function

$$z = -(\xi - p)^2 .$$

The cut along the real axis is not an obstacle for the flow of the liquid directed along x-axis, so the complex potential of the flow is $v_0 z$, where v_0 is the velocity of the flow. The Cartesian net of the plane z is the net of equipotentials and streamlines for this flow.

The inverse mapping of the plane z with the cut onto the domain G is carried out by the composition of functions

$$\xi = i\sqrt{z} + p , \qquad (2.14.1)$$

$$w = f(\xi) ,$$

where the function $f(\xi) = F^{-1}$ is inverse to the function $\xi = F(w)$. The function (2.14.1) maps the Cartesian net onto the net of equipotentials and streamlines of the considered flow with the branching. With the help of formulae (2.14.1) it is possible to visualize the velocity field in streaming around any curve from Catalog 3 with the branching. Note that for the

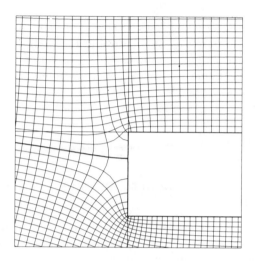

FIGURE 2.8

angular domains with the apex angle less than 2π it is also possible to construct the flow with branching using formulae (2.14.1) but in this case the velocity of the flow at the point at infinity will be infinite. In such domains the flow with branching is not physically realizable.

Example 2.7
The mapping of the Cartesian net carried out by the composition of functions

$$r = i\sqrt{z} + p\,,$$
$$s = i\sqrt{1 - r^2}\,,$$
$$w = rs - \ln(r + s)$$

with $p = 0.3$ is presented in Fig. 2.8. The position of the branching point of the flow can be changed by varying the parameter p. Fig. 2.8 also visualizes the harmonic conjugate field $\mathbf{E} = \mathbf{v}^*$. The field $\mathbf{E}$ can be interpreted as the electric field near the edge of the charged conductive semi-infinite plate placed into the homogeneous electric field.

2.15   Irrotational Streaming Around a Finite Contour

Let a domain G be an exterior of the finite contour L. Designate some constant vector having an angle α with respect to x-axis as $\mathbf{v}_0$. The velocity field of the ideal liquid under irrotational streaming around the contour L is a harmonic plane vector field $\mathbf{v}$ satisfying the conditions

$$\mathbf{v}|_{\infty} = \mathbf{v}_0 \,, \tag{2.15.1}$$

$$v_n|_L = 0 \,, \tag{2.15.2}$$

$$\Gamma = \int_L (\mathbf{v} \cdot \mathbf{t}) \, ds = 0 \,, \tag{2.15.3}$$

where v_n is the normal component of the velocity: $v_n = (\mathbf{v} \cdot \mathbf{n})$ and $(\mathbf{v} \cdot \mathbf{t})$ is the tangential component.

The vector field defined by the conditions (2.15.1)–(2.15.3) has another physical interpretation: it is the magnetic field in the neighborhood of a superconductive cylinder L placed into a homogeneous external magnetic field $\mathbf{v}_0$. The condition (2.15.3) means in this case that the electric current does not flow along the cylinder.

The vector field $\mathbf{v}$ can be visualized by constructing the conformal mapping of the domain G onto the exterior of a segment of the real axis. It can be built as follows.

The domain G can be mapped by some function $\xi = F(w)$ onto the exterior of the unit circle $|\xi| > 1$ so that $F(\infty) = \infty$, $F'(\infty) > 0$. The direction of the flow at infinity is conserved. The flow directed along the x-axis is obtained by the turn $s = e^{-i\alpha}\xi$. The Zhukovskii function $z = 1/2(s + 1/s)$ maps the considered domain onto the exterior of the segment $[-1, 1]$. The segment of the real axis has no influence on the flow directed along the real axis. So the Cartesian net is the net of equipotentials and streamlines of the considered flow in the plane z.

The inverse mapping of the plane z with the cut along the segment $[-1, 1]$ onto the domain G is carried out by the sequence if transformations

$$s = z + \sqrt{z^2 - 1} \,,$$

$$\xi = e^{i\alpha} s \,, \tag{2.15.4}$$

$$w = f(\xi) \,,$$

where $f(\xi) = F^{-1}$ is the function inverse to the function $\xi = F(w)$.

With the help of the formulae (2.15.4) it is possible to visualize the velocity field in the irrotational streaming around any curve presented in the Catalog 2 of the program CONFORM.

Example 2.8

In the Fig. 2.9 the picture of irrotational streaming around the three-stopping epicycloid ($f(\xi) = \xi + 0.5/\xi^2$) is presented when $\alpha = 0.5$. This curve sometimes is called the *Steiner's curve*. The direction of the flow can be changed by varying the parameter α.

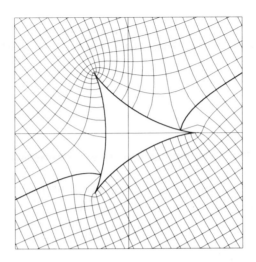

FIGURE 2.9

Fig. 2.9 also visualizes the harmonic conjugate field $\mathbf{A} = \mathbf{v}^*$. The field $\mathbf{A}$ can be interpreted as electrostatic field of a non-charged conductive cylinder L placed into a homogeneous external electric field $\mathbf{A}_0 = [\mathbf{k} \times \mathbf{v}_0]$. Otherwise it can be interpreted as the magnetic field in the neighborhood of the cylinder L made of ideal magnetic material and placed into a homogeneous magnetic field $\mathbf{A}_0$.

In conclusion note that the complex potential of the flow directed along the real axis and streaming around the circumference of the unit radius without circulation is the Zhukovskii function. The isothermic net of the Zhukovskii function depicted above in Fig. 1.21 is the net of equipotentials and streamlines for this flow.

2.16 Mixed Boundary-Value Problem in a Curvilinear Angular Domain. The Simplest Problem of the Filtration Theory

Let G be a curvilinear angular domain of the complex plane w with the boundary divided by the points w_1 and w_2 into three parts L_1, L_2 and L_3, the parts L_1 and L_3 being infinite. Consider the problem on the distribution of the electric current $\mathbf{J}$ in the domain G in the case when the part L_2 is electrically isolated and the parts L_1 and L_3 are equipotentials with the given values of potential U_1 and U_3.

The harmonic vector field $\mathbf{J}$ has the potential $U(u,v)$ being the solution of the mixed boundary value problem for the Laplace equation:

$$\nabla^2 U = 0\,,$$

$$U|_{L_1} = U_1\,,$$

$$\left.\frac{\partial U}{\partial n}\right|_{L_2} = 0\,, \qquad\qquad (2.16.1)$$

$$U|_{L_3} = U_3\,.$$

In particular, the simplest plane problem on the filtration of the ground water under a dam is reduced to the boundary-value problem (2.16.1). In this case the domain G is a curvilinear half-plane (waterproof layer is absent), the curve L_2 is an impenetrable surface of the construction (flutbett) and the curves L_1 and L_3 are isobars — rays of horizontal straight lines: L_1 is the bottom of the upstream wall of the dam and L_3 is the bottom of the downstream wall.

Along with the harmonic field $\mathbf{J} = \nabla U$ we will consider the harmonic conjugate field $\mathbf{A} = \mathbf{J}^*$. The vector field $\mathbf{A}$ is the density of the electric current in the domain G in the case when the part L_2 of its boundary is an electrode supplying the plate by the current $I = U_2 - U_1$. The other electrode is supposed to be placed at the point at infinity and the parts L_1 and L_3 are electrically isolated (they represent current lines for the field $\mathbf{A}$).

The fields $\mathbf{J}$ and $\mathbf{A}$ can be simultaneously visualized by the construction of the conformal mapping of the domain G onto the adequate domain — the half-strip $0 < \operatorname{Im} z < \pi$, $\operatorname{Re} z > 0$. It is necessary to map the domain G conformally onto the half-strip so that the point $w = \infty$ would be mapped into the point $z = \infty$ and the points w_1 and w_2 — into the points $z_1 = i\pi$ and $z_2 = 0$ respectively. The mapping can be constructed as follows. Let the function $t = F(w)$ conformally map the domain G onto the upper half-plane $\operatorname{Im} t > 0$ so that the point $w = \infty$ would be mapped into the point $t = \infty$. Designate $p_1 = F(w_1)$, $p_2 = F(w_2)$. The curve L_2 is mapped onto the segment $[p_1, p_2]$ of the real axis by the function $F(w)$. As the result of linear transformation

$$s = \frac{2t - p_1 - p_2}{p_2 - p_1}$$

the considered domain is mapped onto the upper half-plane $\operatorname{Im} s > 0$, the curve L_2 — onto the segment $[-1, 1]$. Finally, the function

$$z = \operatorname{arch} s$$

maps the half-plane $\operatorname{Im} s > 0$ onto the half-strip $0 < \operatorname{Im} z < \pi$, $\operatorname{Re} z > 0$ with the necessary correspondence of boundaries.

The inverse mapping of the half-plane onto the domain G is carried out

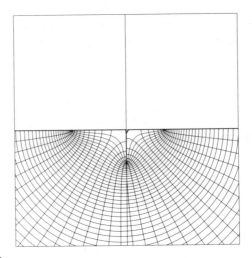

FIGURE 2.10

by the sequence of transformations

$$s = \cosh z \,,$$

$$t = \frac{s(p_2 - p_1) + p_1 + p_2}{2} \,, \qquad (2.16.2)$$

$$w = f(t) \,,$$

where $f(t) = F^{-1}$ is the function inverse to the function $t = F(w)$. The function (2.16.2) maps the Cartesian net in the half-strip $0 < \operatorname{Im} z < \pi$, $\operatorname{Re} z > 0$ onto the net of equipotentials and streamlines for harmonic vector fields $\mathbf{J}$ and $\mathbf{A}$. The solution of the mixed boundary-value problem for any domain presented in Catalog 3 can be visualized with the help of the function (2.16.2). The position of the boundary part L_2 can be changed by varying the real parameters p_1 and p_2 $(p_2 > p_1)$.

Example 2.9

The mapping of the Cartesian net in the half-strip $0 < \operatorname{Im} z < \pi$, $\operatorname{Re} z > 0$ realized by the function (2.16.2) with $f(t) = -i\sqrt{1 - t^2}$, $p_1 = -1.5$, $p_2 = 2$ is presented in Fig. 2.10. The net of lines in Fig. 2.10 can be interpreted as the picture of ground water motion under a dam. The waterproof foundation of the dam is some segment of the real axis (flutbett) together with the cut along the segment of the imaginary axis $[0, -i]$ (cutoff wall). The images of segments of straight lines $\operatorname{Re} z = $ const, $0 < \operatorname{Im} z < \pi$ are streamlines and the images of rays $\operatorname{Im} z = $ const, $\operatorname{Re} z > 0$ are the lines of constant pressure (isobars).

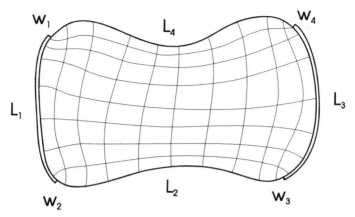

FIGURE 2.11

2.17   Distribution of the Electric Current in a Plate in the Presence of Two Non-Point Electrodes at Its Boundary

Let a homogeneous electrically conductive plate be the singly connected domain G with the boundary L divided by the points w_1, w_2, w_3, w_4 into four parts L_1, L_2, L_3, L_4 numbered in the order of the positive sense of boundary traversal. Consider the problem on the distribution of the electric current in the domain G, assuming that the parts of the boundary L_1 and L_3 (electrodes) have given potentials U_1 and U_3 and the parts L_2 and L_4 are electrically isolated (Fig. 2.11). The distribution of the electric potential in the domain G is the solution of the mixed boundary-value problem for the Laplace equation:

$$\nabla^2 U = 0\,,$$

$$U|_{L_1} = U_1\,,$$

$$U|_{L_3} = U_3\,, \hspace{3cm} (2.17.1)$$

$$\left.\frac{\partial U}{\partial n}\right|_{L_2} = \left.\frac{\partial U}{\partial n}\right|_{L_4} = 0\,.$$

The problem on filtration of the underground water under a dam in the presence of the waterproof layer (Fig. 2.1) is reduced to the problem (2.17.1). Here the part L_1 represents a horizontal half-straight line — the bottom of the downstream wall of the dam, L_2 is a waterproof foundation of the dam (flutbett), L_3 is the bottom of the upstream wall, L_4 — the waterproof layer of the ground.

The problem (2.17.1) can be solved by the conformal mapping of the domain G onto the rectangle. The parts L_1 and L_3 of the boundary are mapped onto the one pair of opposite sides of the rectangle and the parts L_2 and L_4 — onto the other pair. In the process the net of equipotentials and streamlines of the problem (2.17.1) is mapped onto the Cartesian net of lines in the rectangle.

The mapping of the domain G onto the rectangle can be built as follows. Let the function $t = F(w)$ conformally map the domain G onto the upper half-plane $\text{Im } t > 0$. Let the images of the points w_1, w_2, w_3, w_4 be the points of the real axis $t = a, b, c, d$. Without loss of generality we can consider $a < b < c < d$ (otherwise it is possible to change the numbering of the points w_1, w_2, w_3, w_4). The linear-fractional transformation

$$\xi = \frac{\alpha t + \beta}{\gamma t + \delta}$$

(where $\alpha\delta - \beta\gamma > 0$) can map the half-plane $\text{Im } t > 0$ onto the half-plane $\text{Im } \xi > 0$ so that the points a, b, c, d would be mapped into symmetrically placed points $\xi = -1/k, -1, 1, 1/k$ correspondingly. The parameter k is determined by the condition of invariance of the anharmonic ratio D of four points under the linear-fractional mapping:

$$\frac{k}{4}\left(1 + \frac{1}{k}\right)^2 = D, \qquad (2.17.2)$$

where*

$$D = \frac{c-a}{c-b} : \frac{d-a}{d-b}, \qquad D > 1.$$

The solution of the equation (2.17.2) satisfying the condition $0 < k < 1$ is

$$k = \left(\sqrt{D} - \sqrt{D-1}\right)^2.$$

The half-plane $\text{Im } \xi > 0$ is mapped onto the rectangle $-K < \text{Re } z < K$, $0 < \text{Im } z < K'$ by the function $z = F(\xi, m)$, where $m = k^2$.

The inverse mapping of the rectangle onto the domain G is carried out by the sequence of transformations

$$\xi = \text{sn}\,(z, m),$$

$$t = = \frac{\delta\xi - \beta}{-\gamma\xi + \alpha}, \qquad (2.17.3)$$

$$w = f(t),$$

---

*If $d = \infty$ then $D = (c-a)/(c-b)$.

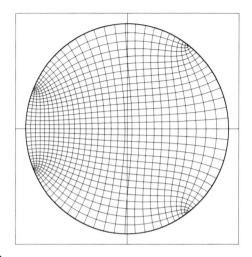

FIGURE 2.12

where $f(t) = F^{-1}$ is the function inverse to the function $t = F(w)$. With the help of (2.17.3) it is possible to visualize the solution of the considered problem for any domain presented in the Catalog 3 of the program CONFORM. For the domains included into other catalogs it is also possible to visualize the solution of the considered problem if you preliminarily map the adequate canonical domain onto a half-plane.

Example 2.10
The distribution of the electric current in the circular plate with two electrodes on its boundary symmetrical with respect to the real axis is shown in Fig. 2.12. The figure represents the mapping of the Cartesian net in the rectangle $-K < x < K$, $0 < y < K'$ carried out by the composition of functions

$$s = \text{sn}\,(z, m)\,,$$
$$w = \frac{p + is}{p - is}$$

when $m = 0.01$ ($K = 1.575$, $K' = 3.696$), $p = 2$.

Example 2.11
The net of streamlines and equipotentials (isobars) is presented in Fig. 2.13 for the problem on the filtration of ground water under a dam in the pres-

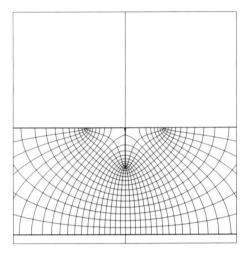

FIGURE 2.13

ence of the waterproof layer. The foundation of the dam is the segment of
the real axis (the flutbett) together with the cut along the segment of the
imaginary axis (the cutoff wall). The figure is created as the mapping of the
Cartesian net in the rectangle carried out by the composition of functions

$$s = \text{sn}(z, m),$$

$$u = \frac{1 + s}{1 - s},$$

$$r = \frac{1}{2}\left(u + \frac{1}{u}\right),$$

$$v = r(1 - p) + p,$$

$$t = v + i\sqrt{1 - v^2},$$

$$w = \ln t$$

with $m = 0.5$ ($K = K' = 1.85$), $p = 0.3$.

2.18 The General Problem on the Streaming Around a Finite Contour

Let a domain $\overset{\circ}{G}$ of the w plane be an exterior of a contour L. The general
problem on the streaming around the contour L is the problem of determining the harmonic vector field, which has the given circulation Γ and the

velocity $\mathbf{v}_0$ at the point at infinity:

$$[\nabla \times \mathbf{v}] = 0, \quad (\nabla \cdot \mathbf{v}) = 0,$$

$$\mathbf{v}|_\infty = \mathbf{v}_0,$$

$$v_n|_L = 0, \tag{2.18.1}$$

$$\Gamma = \int_L (\mathbf{v} \cdot \mathbf{t}) \, ds \neq 0.$$

The field $\mathbf{v}$ is determined uniquely by the conditions (2.18.1). Together with hydrodynamics the vector field $\mathbf{v}$ has another physical interpretation. It can be treated as a magnetic field in the neighborhood of superconductive cylinder L with the electric current Γ flowing along it in the presence of the homogeneous external magnetic field $\mathbf{v}_0$.

Harmonic conjugate vector field $\mathbf{E} = \mathbf{v}^*$ has the meaning of electrostatic field of conductive cylinder L placed into the homogeneous external electric field $\mathbf{E}_0 = [\mathbf{k} \times \mathbf{v}_0]$ and carrying the electric charge Γ per unit length.

The motion of a liquid with circulation can be treated as the superposition of two motions — the purely rotational motion with circulation Γ and zero velocity at infinity and the irrotational motion. First consider the general problem on streaming around a circular cylinder.

2.18.1 Streaming around a circular cylinder

Let a domain D of the complex plane z be an exterior of the circle with the radius R: $|z| > R$. The complex potential of purely rotational streaming around the circle with circulation Γ is the infinitely-valued function

$$\Phi_1 = \frac{\Gamma}{2\pi i} \operatorname{Ln} z.$$

To calculate the complex potential of irrotational streaming around the circle by the flow with the velocity $v_0 = |v_0| e^{i\alpha}$ first we make a turn: $\zeta = e^{-i\alpha} z$. In the plane ζ we have the problem on irrotational streaming around the circle of radius R by a flow having at infinity the velocity $|v_0|$ directed along the real axis. With the help of the function

$$\varphi(\zeta) = \zeta + \frac{R^2}{\zeta}$$

the exterior of the circle is conformally mapped onto the exterior of a segment of the real axis $(-2R, 2R)$ so that $\varphi(\infty) = \infty$, $\varphi'(\infty) = 1$.*

The segment of the real axis does not disturb the flow directed along the real axis. The complex potential of the flow equals to $|v_0| \varphi$. Returning to

---

*The latest condition is necessary for not to recalculate the velocity of the flow at the point at infinity.

z-variable we write the complex potential as

$$\Phi_2 = |v_0| \left(e^{-i\alpha} z + \frac{R^2 e^{i\alpha}}{z} \right) = \bar{v}_0 \, z + v_0 \frac{R_2}{z} \, .$$

The complex potential of rotational streaming around a circle is

$$\Phi = \Phi_1 + \Phi_2 = \bar{v}_0 z + v_0 \frac{R^2}{z} + \frac{\Gamma}{2\pi i} \, \text{Ln} \, z \, . \qquad (2.18.2)$$

The net of streamlines and equipotentials of the flow (2.18.1) is the isothermic net of the function (2.18.2). This net of lines can be built by the CONFORM program.

The complex potential of rotational streaming around a circle has different appearance for different values of Γ. The velocity of the flow is determined in accordance with (2.3.3) as

$$\overline{v(z)} = \Phi'(z) = \bar{v}_0 - v_0 \frac{R^2}{z^2} + \frac{\Gamma}{2\pi i z} \, .$$

The critical points of the flow are the roots of the quadratic equation

$$\bar{v}_0 z^2 + \frac{\Gamma}{2\pi i} z - v_0 R^2 = 0 \, . \qquad (2.18.3)$$

In the case when v_0 is real-valued ($v_0 = |v_0| > 0$) the equation (2.18.3) has the roots

$$z_{1,2} = -\frac{\Gamma}{4\pi i v_0} \pm \sqrt{R^2 - \frac{\Gamma^2}{16\pi^2 v_0^2}} \, .$$

If $|\Gamma| < 4\pi R v_0$ both of the roots of the equation (2.18.3) are complex-valued, $|z_{1,2}| = R$. In this case one of the critical points is the branch point of the flow and the other — the convergence point. When $|\Gamma| < 4\pi R v_0$ the circumference $|z| = R$ is a part of the branching streamline. The figure of streamlines and equipotentials in this case is shown in Fig. 2.14, a. If $|\Gamma| > 4\pi R v_0$ both of the critical points $z_{1,2}$ are purely imaginary:

$$z_{1,2} = -\frac{\Gamma}{4\pi i v_0} \pm i \sqrt{\frac{\Gamma^2}{16\pi^2 v_0^2} - R^2} \, .$$

One of the points $z_{1,2}$ lies outside the circle, the other — inside the circle $|z| < R$, i.e., outside the flow region. In this case there exists a domain near the circle, where the trajectories of particles of the liquid are closed. The figure of streamlines and equipotentials of this case is shown in Fig. 2.14, b.

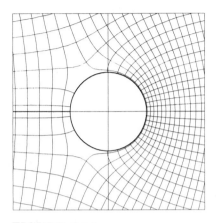
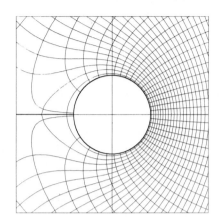

FIGURE 2.14

2.18.2 Rotational streaming around an arbitrary cylinder

The problem on streaming around a cylinder of an arbitrary cross-section (2.18.1) can be reduced to the problem of streaming around a circular cylinder with the help of a conformal mapping.

Let the function $z = F(w)$ map the exterior of the contour L onto the exterior of the unit circle $|z| > 1$ so that $F(\infty) = \infty$, $F'(\infty) > 0$. The function $F(w)$ has a first order pole at the point at infinity: $F(w) = Cw + \psi(w)$, where $\psi(w)$ is the function regular at the point at infinity, $C = F'(\infty) > 0$. The function $\xi = \varphi(w) = F(w)/C$ maps the domain G onto the exterior of the circle $|\xi| > R$ (where $R = 1/C = 1/F'(\infty)$) so that $\varphi(\infty) = \infty$, $\varphi'(\infty) = 1$.[†]

In the plane ξ we have the problem on streaming around the circle of the radius R by the flow with the velocity $v_0 = |v_0|e^{i\alpha}$ and circulation Γ. Its solution is expressed by the formula (2.18.2):

$$\Phi = |v_0| \left(e^{-i\alpha}\xi + \frac{e^{i\alpha}R^2}{\xi} \right) - iQ \ln \xi \,,$$

where $Q = \Gamma/2\pi$.

Returning to the variable w and omitting unessential constant term we obtain the complex potential of streaming around an arbitrary contour L as

$$\Phi = \frac{|v_0|}{F'(\infty)} \left(e^{-i\alpha}F(w) + \frac{e^{i\alpha}}{F(w)} \right) - iQ \ln F(w) \,. \qquad (2.18.4)$$

[†]The value $R = 1/C$ is called the *conformal radius* of the domain G.

To visualize the flow with circulation (2.18.1) the **Flow** mode is present in the program CONFORM. In this mode the program plots the isothermic net of the function (2.18.4) with $|v_0| = F'(\infty)$.

The function (2.18.4) is given in the parametric form

$$\Phi = e^{-i\alpha}z + \frac{e^{i\alpha}}{z} - iQ \ln z,$$

$$w = f(z), \qquad\qquad\qquad (2.18.5)$$

where $f(z) = F^{-1}$ is the function inverse to the function $z = F(w)$ and z is the parameter accepting arbitrary values from the domain $|z| > 1$. Thus the program CONFORM permits the visualization of a rotational streaming around any contour included into Catalog 2. The mapping function (user-defined or loaded from Catalog 2) and two real parameters: an angle of attack α and normalized circulation Q should be entered into the program. If $|Q| < 2$ a figure with two critical points on the curve L is obtained. If $|Q| > 2$ the flow has one critical point outside the curve L. In this case a domain filled with closed trajectories exists in the neighborhood of the curve L.

Example 2.12

A net of streamlines and equipotentials in the streaming around circular lune (lens) created with the help of the mapping function

$$s = \left(\frac{z-1}{z+1}\right)^{2p},$$

$$w = \frac{1+s}{1-s}$$

with $p = 0.8$, $\alpha = 0.6$, $Q = 1$ is shown in Fig. 2.15.

Equipotentials shown in Fig. 2.15 have breaks in the negative real half-axis. This is connected with an ambiguity of the complex potential (2.18.4). Extracting the single-valued branch of the function $\mathrm{Ln}\, z$ one obtains the function with discontinuity in the negative real half-axis. However in the reality neither velocity field, nor equipotentials have any break.

Note in the conclusion that the formulae (2.18.4)–(2.18.5) are valid in the frame of reference where the cylinder is at rest and the liquid is moving. Then the term

$$\Phi_0 = \bar{v}_0 z = \bar{v}_0 F(w)$$

in the expression (2.18.4) is the potential of nonperturbed motion of the liquid. If the liquid is at rest at infinity and the cylinder is moving with the velocity $-\mathbf{v}_0$ then the term Φ_0 is absent in the expression (2.18.4). In

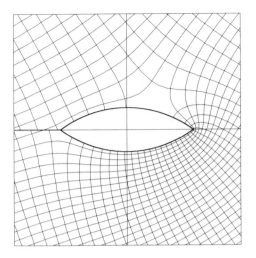

FIGURE 2.15

this frame of reference associated with the liquid the complex potential of the rotational motion is equal to

$$\Phi = v_0 \frac{R^2}{F(w)} + \frac{\Gamma}{2\pi i} \operatorname{Ln} F(w).$$

The flow can be visualized in this frame of reference by using the parametric representation of the complex potential

$$\Phi = \frac{e^{i\alpha}}{z} - iQ \ln z,$$

$$w = f(z) \tag{2.18.6}$$

by the program CONFORM. The **Isoline** mode should be used and the complex potential (2.18.6) should be entered into the first editor window and the mapping function — into the second one.

2.18.3   Streaming around aerodynamical profiles

In some problems of aerodynamics of small speeds (when calculating the motion of an aircraft with the speed much lower than the speed of sound) the air can be considered as an ideal noncompressible liquid (Subsection 2.2.1). In the process of the calculation of streaming around the aerofoil of a large span the aerofoil can be considered as a long cylinder and the flow of an air — as a plane-parallel one. In this case to calculate the velocity field of air it is possible to use the formulae (2.18.4)–(2.18.5).

Plane curves used to approximate the section of an aerofoil are called *aerodynamical profiles*. As a rule, aerodynamical profiles have sharp edges.

Generally speaking the velocity of air at the sharp edge is infinite under the streaming around the aerodynamical profile by the flow with the given speed and angle of attack[‡] and an arbitrary circulation. However it is always possible to fit the circulation so that the convergence point would coincide with the sharp edge. Such a circulation will be called *preferable*.[§] In the program CONFORM the preferable circulation is determined by the fitting.

The simplest and historically the first curve approximating the aerofoil profile is the Zhukovskii profile, being the mapping of a circumference made with the help of the Zhukovskii function (Subsection 1.7.4). The Zhukovskii profile has the returning point, i.e., an angle at the sharp edge is zero. Such profiles are not technologically realizable. The best approximation of the real profiles is given by the *Carman-Treftz profile*,[114-119] being a mapping of a circumference by the generalized Zhukovskii function (Subsection 1.9.1).

Example 2.13

The figure of the streaming around the Carman-Treftz profile is shown in Fig. 2.16. It was created with the help of the function

$$s = i \tan a - d e^{-i\alpha} \,,$$

$$t = z(1/\cos a + d) + s \,,$$

$$r = \left(\frac{t-1}{t+1} \right)^{2-p} \,,$$

$$w = \frac{1+r}{1-r}$$

with $a = 0.2$, $d = 0.1$, $p = 0.05$.

The figure was created by the program CONFORM in the **Flow** mode. The angle of attack was chosen to be equal to 0.4, the circulation Q was fitted so that the convergence point of the flow was placed at the profile edge. For the given parameters the preferable circulation value appeared to be equal to $Q = -1.11$.

[‡] An *angle of attack* is the angle between the direction of the flow and some of the chords of the profile.

[§] The statement that in the real flow the convergence point is shifted onto the sharp edge under the influence of the finite viscosity and compressibility of the air is called in aerodynamics the *Zhukovskii-Chaplygin postulate*. This postulate permits one to determine the circulation arising in streaming around the given profile and the airlift connected with it.

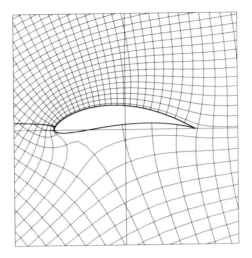

FIGURE 2.16

2.19   The Problem on the Impact of a Solid Cylinder upon the Surface of Non-Compressible Liquid

Let L' be a non-closed finite curve lying in the half-plane Im $w < 0$. Its ends (points A and B) lie on the real axis (Fig. 2.17). Designate as $\mathcal{R}'$ the part of the real axis lying outside the segment $[A, B]$, as G' — the part of the half-plane Im $w < 0$ lying outside the curve L'. Consider the curve L' as the directrix of an infinite solid cylinder also denoted as L'. Assume that the domain G' is filled by the ideal liquid staying at rest before the moment $t = 0$.

Let the cylinder accept at the moment $t = 0$ (for example, after the impact) the speed $\mathbf{V}_0$ directed vertically downwards: $\mathbf{V}_0 = -V_0\,\mathbf{j}$. Consider the problem on the determination of the velocities of liquid particles just after the impact. At the moment of the impact the stressed state arises characterized by the shock (pulse) pressure $p(u, v)$. After the impact a plane harmonic velocity field $\mathbf{V}$ with the potential $U(u, v) = p(u, v)/\rho$ arises in the liquid, ρ is the density of the liquid. On the free surface $\mathcal{R}'$ the pressure p and thus the potential U vanish. On the surface of the cylinder L' the normal component of the velocity V_n is equal to $(\mathbf{V}_0 \cdot \mathbf{n})$, where $\mathbf{n}$ is the unit normal to the curve L'.

Therefore the problem on the determination of the velocity field of the liquid $\mathbf{V} = \nabla U$ just after the impact is reduced to the mixed boundary-value problem for the potential $U(u, v)$:

$$\nabla^2 U = 0 \quad \text{in the domain } G',$$

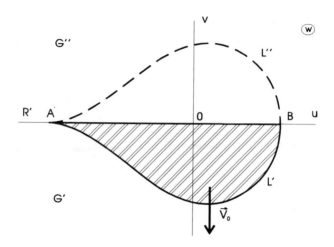

FIGURE 2.17

$$\frac{\partial U}{\partial n}\bigg|_{L'} = (\mathbf{V}_0 \cdot \mathbf{n})\,,$$

$$U|_{\mathcal{R}'} = 0\,, \tag{2.19.1}$$

$$U|_\infty = 0\,.$$

The problem (2.19.1) can be reduced to the problem of the motion of the cylinder in the unbounded medium (Section 2.18) with the help of the symmetry principle.

Consider the domain G'' to be symmetrical to the domain G' with respect to the real axis $\mathcal{R}$ and the curve L'' to be symmetrical to the curve L'. Designate the association of the curves L' and L'' as L and that of the domains G' and G'' — as G. An analytical continuation of the function U from the domain G' into the domain G'' results in the problem on motion of the symmetrical cylinder L with the speed $\mathbf{V}_0$ and zero circulation in the resting unbounded medium. Indeed, the real axis $\mathcal{R}$ is an equipotential in this case. The complex potential of the liquid velocities in this problem was obtained in the Section 2.17. Assuming in the formula (2.18.6) $v_0 = i$, $Q = 0$ we have

$$\Phi = \frac{i}{z} = \frac{i}{F(w)}\,, \tag{2.19.2}$$

where $F(w)$ is the function realizing the conformal mapping of the domain G onto the exterior of the unit circle $|z| > 1$ and satisfying the conditions $F(\infty) = \infty$, $F'(\infty) > 0$. The net of streamlines and equipotentials (shock isobars) is the isothermic net of the function (2.19.2). To visualize the velocity field it is more convenient to represent the dependence of the function

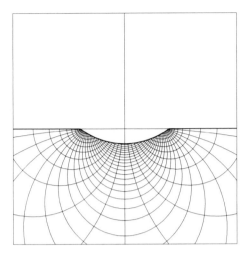

FIGURE 2.18

Φ on the coordinates of the point w in the parametric form

$$\Phi = \frac{i}{z},$$

$$w = f(z),\qquad\qquad (2.19.3)$$

where $f(z) = F^{-1}$ is the function inverse to the function $z = F(w)$ and the complex parameter z accepts any of the values from the domain $\operatorname{Im} z < 0$, $|z| > 1$.

Using the program CONFORM in the mode **Isoline** it is possible with the help of the formulae (2.19.3) to visualize the solution of this problem for any symmetrical (with respect to the real axis) domain included into Catalog 2.

Note, that the motion of the real liquid after the impact is described by the formula (2.19.3) only in the case when the normal to the curve L' forms an acute angle with the vector $\mathbf{V}_0$, i.e., if the liquid is subjected to compression during the impact. If at a part of the curve L' the condition $(\mathbf{V}_0 \cdot \mathbf{n}) > 0$ is violated the motion of the real liquid will be considerably different from the model of the motion without separation, because in this part of the curve L' the separation of the liquid from the wall will occur (cavitation).

Example 2.14

The streamlines and shock isobars under the impact of a segment of a circle upon an incompressible liquid are shown in Fig. 2.18.

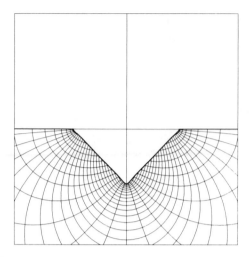

FIGURE 2.19

Example 2.15

The streamlines and shock isobars under the impact of a wedge with the right angle upon an incompressible liquid are shown in Fig. 2.19.

2.20   The Flow with Branching in a Curvilinear Strip with N Branches

Consider a curvilinear strip G with N branches bounded by infinite curves $L_1, L_2, \ldots, L_N$. Designate branches of the strip as $A_1, A_2, \ldots, A_N$. Let sources (sinks) of intensity p_k be placed in the branches of the strip A_k and $\sum_1^N p_k = 0$. It is necessary to construct the harmonic vector velocity field $\mathbf{V}$ of the ideal liquid in the domain G assuming that the curves L_k are the streamlines. In the case when more than two numbers p_k differs from zero, let us call the vector field a flow with branching.

Along with the velocity field $\mathbf{V}$ we will consider harmonic conjugate vector field $\mathbf{E} = \mathbf{V}^*$, the curve L_k being equipotentials. The field $\mathbf{E}$ represents the plane electrostatic field in the domain G bounded by N infinite electrodes. The values p_k are potential differences between the curves L_k and L_{k-1}. If the potential of the curve L_N is equal to U_N, the expression

$$U_m = U_N + \sum_{k=1}^{m} p_k \, .$$

is obtained for potential of the curve L_m.

Thus the considered hydrodynamic problem for the field $\mathbf{V}$ is equivalent to the electrostatic problem for the field $\mathbf{E}$, which is reduced to the Dirichlet problem for the potential of the electrostatic field $U(u, v)$ with the constant values of the potential on the curves L_k:

$$\nabla^2 U = 0 \quad \text{in the domain } G,$$

$$U|_{L_k} = U_k.$$

The complex potential of the vector field $\mathbf{V}$ can be constructed by the conformal mapping of the domain G onto the half-plane Im $t > 0$. Let such a mapping be realized by the function $t = F(w)$ and let the images of the branches A_k of the strip be points of the real axis $t = a_k$ $(k = 1, 2, \ldots, N)$. Without loss of generality we can consider $a_1 < a_2 < \ldots < a_N$. If all the points a_k are finite, the complex potential of the field $\mathbf{V}$ will be equal to

$$\Phi = \frac{1}{\pi} \sum_{k=1}^{N} p_k \ln(t - a_k). \tag{2.20.1}$$

If one of the points a_k (namely a_N) is a point at infinity, the corresponding term vanishes in the formula (2.20.1):

$$\Phi = \frac{1}{\pi} \sum_{k=1}^{N-1} p_k \ln(t - a_k). \tag{2.20.2}$$

Substituting the function $t = F(w)$ into the formula (2.20.1) or (2.20.2) we obtain the complex potential Φ as the function of the coordinates of the point w. However it is more convenient to build the isothermic net of the function $\Phi(w)$ specifying it in the parametric form:

$$\Phi = \frac{1}{\pi} \sum_{k=1}^{N} p_k \ln(t - a_k),$$

$$w = f(t), \tag{2.20.3}$$

where $f(t) = F^{-1}$ is the function inverse to the function $t = F(w)$. Using the program CONFORM in the **Isoline** mode it is possible to build streamlines and equipotentials for any domain presented in Catalog 5 by formulae (2.20.3).

Example 2.16
The net of equipotentials and force lines in the plane with four symmetrical star-shaped cuts is shown in Fig. 2.20, a, b. The **Isoline** mode of the program CONFORM was used and the dependence of the complex potential Φ on coordinates of the point w was given in the parametric form. The coordinates of the point w were obtained under the conformal mapping of

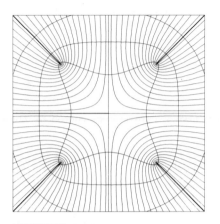 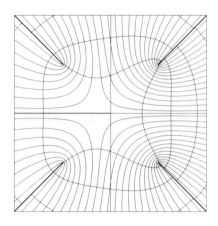

FIGURE 2.20

the circle $|z| < 1$ onto the considered domain with the help of the function $w = z/\sqrt{1 - z^4}$. The points $z = \pm 1$, $z = \pm i$ are mapped into the branches of the curvilinear strip (Section 2.13). The mapping of the circle $|z| < 1$ onto the half-plane $\text{Im}\, t > 0$ is carried out by the function

$$t = i\,\frac{z - 1}{z + 1},$$

preimages of the branches are mapped into the points $t_1 = -1$, $t_2 = 0$, $t_3 = 1$, $t_4 = \infty$. The complex potential of the flow created by the sources p_k in the branches of the strip is expressed by the formula (2.20.2):

$$\Phi = \frac{1}{\pi}\left[p_1 \ln(t + 1) + p_2 \ln t + p_3 \ln(t - 1)\right].$$

Fig. 2.20, a was created with $p_1 = 1$, $p_2 = -1$, $p_3 = 1$, and $p_4 = -1$.

In electronic optics the device with four symmetrically placed electrodes having potentials with alternating signs is called a *quadrupole lens*).[137, 146] Therefore, the net of lines in Fig. 2.20, a visualizes the electrostatic field of the quadrupole lens.

Fig. 2.20, b is created with $p_1 = 1$, $p_2 = -1.3$, $p_3 = 1$, $p_4 = -0.7$. It represents the electrostatic field in quadrupole lens with potentials of electrodes mismatched. Note, that the latest image can be created only with the use of **Isoline** mode of the program CONFORM.

3

Practice: Catalog of Conformal Mappings

This Chapter contains a catalog of the mappings carried out by elementary and elliptic functions. The catalog includes only the mappings of single-connected domains. The mappings of doubly-connected domains considered in the Section 1.14 are not included here to avoid repetitions.

The catalog consists of five parts corresponding to classification of single-connected domains:

1. Finite domains.

2. Exteriors of finite contours.

3. Curvilinear angular domains.

4. Curvilinear strips.

5. Curvilinear strips with three or four branches.

In the first column the function $w = f(z)$ with concrete values of parameters is shown. The second column contains the acceptable range of parameters and some comments. The third column presents the visualization of the conformal mapping obtained with the help of the program CONFORM.

3.1 Catalog 1. Finite domains

In this catalog for each finite domain the function $w = f(z)$ which maps the circle $|z| < 1$ onto this domain is presented. For domains symmetrical with respect to the origin the function $f(z)$ satisfies the conditions $f(0) = 0$, $f'(0) > 0$. Such a function is determined uniquely.

#	Function	Comment	Figure
1	$p = 1.5$ $s = i\dfrac{1+z}{1-z}$ $t = -s + i\sqrt{1-s^2}$ $w = t^p$	$0 < p < 2$, sector of a circle	
2	$p = 1.2$ $r = i\dfrac{1+z}{1-z}$ $s = p - r^2$ $t = s - \sqrt{s-1}\sqrt{s+1}$ $a = 0.6$ $w = t^a$	$0 < a < 1$, $p > 1$, sector with a cut	
3	$p = 4$ $s = \ln\dfrac{1+z}{1-z}$ $t = \dfrac{s(p-1)}{i\pi} + \dfrac{p+1}{2}$ $w = \dfrac{1}{t}$	$p > 1$, lune between two tangent circumferences	
4	$a = 0.1$ $p = 1.5$ $r = \dfrac{z + i\sin a}{\cos a}$ $s = \dfrac{r+1}{1-r}$ $t = s^p$ $w = \dfrac{t-1}{1+t}$	$|a| < \pi/2$, $0 < p < 2$, finite circular lune	

#	Function	Comment	Figure
5	$p_1 = 1.193$ $p_2 = 1.493$ $s = \dfrac{1}{2}\left(z + \dfrac{1}{z}\right)$ $t = \dfrac{1}{2}\left[s(p_1+p_2)+p_2-p_1\right]$ $w = t - \sqrt{t-1}\sqrt{t+1}$	$p_1, p_2 \geq 1$, circle with two cuts	
6	$n = 3$ $p = 1.2029$ $r = \dfrac{p}{2}\left(1 + z^n\right)$ $t = r + \sqrt{r^2 - z^n}$ $w = z\, t^{-2/n}$	$n > 1$, $p > 1$, integer n, circle with n symmetrical cuts	
7	$p = 1$ $t = z/p$ $w = (1+t)^2$	$p = 1$ — cardioid, $p > 1$ — limaçon	
8	$p = 1.005$ $t = z/p$ $w = \sqrt{1+t}$	$p = 1$ — a loop of Bernoulli's lemniscata, $p > 1$ — a loop of two-contour cassinian	

#	Function	Comment	Figure
9	$p = 0.8$ $w = (1 + z)^p$	$1 < p < 2$ — generalized cardioid, $0 < p < 1$ — a loop of a generalized lemniscata	
10	$p = 1.2$ $t = z^2$ $s = \dfrac{1 - p\,t}{t - p}$ $r = p\,s$ $w = \sqrt{r + 1}\,\operatorname{sign}(\operatorname{Re} z)$	$p > 1$, single-contour Cassini's oval	
11	$n = 3$ $p = 1.0$ $s = z/p$ $w = s - \dfrac{s^{n+1}}{n + 1}$	integer n, $p = 1$ — epicycloid, $p > 1$ — shortened epitrochoid	
12	$p = 1.5$ $s = z/p$ $w = \dfrac{2s}{s^2 + 1}$	$p > 1$, elliptic Booth's lemniscata	

#	Function	Comment	Figure
13	$m = 0.794$ $t = \dfrac{2z}{z^2 + 1}$ $w = F(t, m)$	$0 < m < 1$, rectangle	
14	$m = 0.794$ $p = 0.951$ $t = \dfrac{2z}{z^2 + 1}$ $w = F(pt, m)$	$0 < m < 1$, $0 < p < 1$, rectangle with two cuts	
15	$m = 0.0165$ $t = \dfrac{2z}{z^2 + 1}$ $u = \dfrac{\pi}{2K'(m)} F(t, m)$ $w = e^u$	half of a circular annulus	
16	$r = \dfrac{2z}{z^2 + 1}$ $m = 0.95$ $s = r/\sqrt{m}$ $t = \dfrac{\pi}{2K(m)} F(s, m)$ $w = \sin t$	$0 < m < 1$, ellipse	

#	Function	Comment	Figure
17	$m = 0.9463$ $s = -i\dfrac{z+1}{z-1}$ $t = \dfrac{\pi}{2K(m)} F(s,m)$ $w = \sin t$	$0 < m < 1$, half of an ellipse	
18	$m = 0.0294372$ $p = \sqrt{m}$ $v = \dfrac{p+1}{2}\left(z + \dfrac{1}{z}\right)$ $u = -\dfrac{v+3-p}{v-1+3p}$ $s = F(u,m)$ $t = \dfrac{1}{2}\left(\dfrac{s}{K(m)} + 1\right)$ $w = i\,t^2$	domain between two arcs of confocal parabolas	
19	$t = i\dfrac{1+z}{1-z}$ $s = t - i\sqrt{1-t^2}$ $x = (1+i)/\sqrt{2}$ $r = s\,x$ $u = \dfrac{2r}{1+r^2}$ $w = i\,x\,F(u, 0.5)$	right-angle triangle	
20	$m = 0.962$ $s = \dfrac{2z}{1+z^2}$ $t = \dfrac{\pi}{2K(m)} F(s,m)$ $w = \sin t$	$0 < m < 1$, ellipse with two cuts	

3.2   Catalog 2. Exteriors of finite contours

In this catalog for each domain the function $w = f(z)$ maps the exterior of the circle $|z| > 1$ onto the considered domain. The function $f(z)$ satisfies the conditions $f(\infty) = \infty$, $f'(\infty) > 0$. These conditions determine the function $f(z)$ uniquely.

#	Function	Comment	Figure
1	$a = 1$ $b = 0.5$ $w = \dfrac{a+b}{2}\,z + \dfrac{a-b}{2z}$	ellipse with half-axis a and b	
2	$a = 1$ $b = 2$ $p_1 = 1.207$ $p_2 = 1.207$ $r = \frac{1}{2}\left(z + \frac{1}{z}\right)$ $s = 0.5[(p_1 + p_2)\,r + p_2 - p_1]$ $t = s + \sqrt{s-1}\sqrt{s+1}$ $w = \dfrac{a+b}{2}\,t + \dfrac{a-b}{2t}$	Exterior of an ellipse with two cuts	
3	$p = 1.5$ $r_0 = e^{i\pi p/(1+p)}$ $r = \dfrac{z - 1/r_0}{z - r_0}$ $s = \ln r$ $w = \dfrac{\pi i}{s(1+p)}$	two tangent circles	
4	$a = 0.3$ $p = 0.5$ $r = i\,\dfrac{z + \sin a}{\cos a}$ $s = \dfrac{r-1}{r+1}$ $t = s^p$ $w = -i\,\dfrac{1+t}{1-t}$	$|a| < \pi/2$, $0 < p \le 2$, circular lune	

#	Function	Comment	Figure
5	$p_1 = 1.4172$ $p_2 = 2.2376$ $s = \dfrac{1}{2}\left(z + \dfrac{1}{z}\right)$ $t = 0.5[s(p_1 + p_2) + p_1 - p_2]$ $w = t\sqrt{1 - t^{-2}}$	$p_1, p_2 \geq 1$, cross-like cut	
6	$p = 1.7989$ $s = \dfrac{p}{2}\left(z + \dfrac{1}{z}\right)$ $t = \dfrac{p}{p^2 + 1}\left(s + \dfrac{1}{s}\right)$ $w = t - i\sqrt{1 - t^2}\,\text{sign}\,(\text{Im}\,z)$	$p > 1$, two symmetrical arcs of a circumference and its diameter	
7	$n = 3$ $w = z\left(1 + \dfrac{1}{z^n}\right)^{2/n}$	integer $n \geq 3$, star	
8	$n = 3$ $p = 1.2451$ $s = \dfrac{p}{2}\left(1 + \dfrac{1}{z^n}\right)$ $t = s + \sqrt{s^2 - \dfrac{1}{z^n}}$ $w = z\,t^{2/n}$	integer $n > 2$, circle and n symmetrical cuts	

#	Function	Comment	Figure
9	$p = 0.3$ $t = z\,(p+1) - p$ $w = \dfrac{1}{2}\left(t + \dfrac{1}{t}\right)$	$p > 0$, symmetrical Zhukovskii profile	
10	$a = 0.25$ $p = 0.7$ $s = \dfrac{1}{2}\left(z + \dfrac{1}{z}\right)$ $t = \dfrac{s\,(1+p) + 1 - p}{2}$ $w = t\left(1 - \dfrac{1}{t}\right)^{a}$	$p \geq 0$, $0 < a < 1$, symmetrical star-shaped cut with three rays	
11	$s = \dfrac{1}{2}\left(z + \dfrac{1}{z}\right)$ $t = \dfrac{1.0539}{s} + \ln\dfrac{s+1}{s-1}$ $w = \dfrac{2\pi}{t}$	two tangent half- circumfer- ences	
12	$s = \dfrac{1}{2}\left(z + \dfrac{1}{z}\right)$ $t = 0.27847\,\dfrac{s-1}{s+1}$ $w = -\dfrac{i}{t + \ln t + 1}$	two tangent half- circumfer- ences	

#	Function	Comment	Figure
13	$p = 1.509$ $s = 2i\arctan(1/z)$ $t = ps + i\sinh s$ $w = 1/t$	two tangent half-circumfer-ences	
14	$p = 0.662744$ $r = ip\dfrac{z+1}{z-1}$ $s = i\sqrt{1-r^2}$ $t = s - i\ln\dfrac{s+i}{r}$ $w = \pi/t$	two tangent half-circles	
15	$p = 1.83681$ $r = ip\dfrac{z+1}{z-1}$ $s = i\sqrt{1-r^2}$ $t = s - i\ln\dfrac{s+i}{r} - \dfrac{i\pi}{2}$ $w = \pi/t$	two tangent lenses	
16	$p = 1.05$ $t = pz$ $w = t\sqrt{1+\dfrac{1}{t^2}}$	$p = 1$ — lemniscata, $p > 1$ — single-contour cassinian	

#	Function	Comment	Figure
17	$n = 3$ $w = z \left(1 + \dfrac{1}{z^n}\right)^{1/n}$	n-th order lemniscata	
18	$p = 0.17$ $s = \arctan(1/z)$ $t = 4\,p\,s$ $w = 1/\sin t$	$0 < p < \dfrac{1}{2}$, Booth's lemniscata	
19	$p = 1.1$ $n = 2$ $w = z + \dfrac{1}{p\,n\,z^n}$	integer $n \geq 2$, $p \geq 1$, hypotrochoid	
20	$s = \dfrac{1}{\pi} \arccos \dfrac{z-1}{z+1}$ $w = s^{-2} - 1$	cardioid	

#	Function	Comment	Figure
21	$t = \ln \dfrac{z^2 + 1}{z^2 - 1}$ $w = \dfrac{\text{sign}\,(\text{Re}\ z)}{\sqrt{t}}$	four-petal rose consisting of two lemniscatas	
22	$p = 0.5175$ $s_0 =$ $\left(\dfrac{p}{2} + i\sqrt{1 - \dfrac{p^2}{4}} \right)^2$ $s = \dfrac{z\, s_0 - \overline{s_0}}{z - 1}$ $r = \left(\dfrac{1 - \sqrt{s}}{1 + \sqrt{s}} \right)^p$ $w = r\, \dfrac{1 + p\sqrt{s} + s}{1 - p\sqrt{s} + s}$	$0 < p < 2$, sector	
23	$p = 0.6$ $q = E(1 - p)$ $r = K(1 - p)$ $t = \dfrac{1}{2}\left(z + \dfrac{1}{z} \right)$ $s = E(t, 1/p)$ $w = s +$ $i\, \dfrac{q - p\,r}{\sqrt{p}}\, \text{sign}\,(\text{Im}\ t)$	rectangle	
24	$m = 0.998$ $r = \dfrac{2z}{z^2 + 1}$ $s = r/\sqrt{m}$ $t = \dfrac{\pi}{2K(m)}\, F(s, m)$ $w = 1/\sin t$	elliptic Booth's lemniscata	

#	Function	Comment	Figure
25	$m = 0.96$ $k_1 = \sqrt{1 - m}$ $z_0 = (1 - k_1)/\sqrt{m}$ $z_1 = \dfrac{z + z_0}{1 + z\, z_0}$ $s = \dfrac{2z_1}{\sqrt{m}(z_1^2 + 1})$ $t = \dfrac{\pi}{2K(m)}\, F(s, m)$ $w = 1/(\sin t - 1)$	Pascal's limaçon	
26	$m = 0.0955$ $m_1 = 1 - m$ $u = \dfrac{1}{2\sqrt{m}}\left(z + \dfrac{1}{z}\right)$ $a = 1 - \dfrac{E(m_1)}{K(m_1)}$ $w =$ $E(u, m) - a F(u, m)$	$0 < m < 1$, H-shaped cut	
27	$m = 0.5$ $m_1 = 1 - m$ $p = 1/m$ $u = \dfrac{1}{2}\left(z + \dfrac{1}{z}\right)$ $h =$ $\dfrac{E(m_1) - m\, K(m_1)}{\sqrt{m}}$ $t = E(u, p) -$ $i\, h\, \text{sign}\,(\text{Im}\, u)$ $W = -t^2$	parabolic lens	
28	$a = 0.66$ $m = 0.16$ $m_1 = 1 - m$ $t = \dfrac{1}{2\sqrt{m}}\left(z + \dfrac{1}{z}\right)$ $s =$ $E(t, m) - a\, F(t, m)$ $h = E(m_1) -$ $(1 - a)\, K(m_1)$ $w = s - i\, h\, \text{sign}\,(\text{Im}\, t)$	rectangle with cuts	

#	Function	Comment	Figure
29	$m = 0.05$ $r = -\dfrac{2z}{z^2+1}$ $s = \dfrac{\pi\, F(r,m)}{K(1-m)}$ $t = e^s$ $w = \dfrac{1+t}{1-t}$	$0 < m < 1$, "dumb-bells"	
30	$p = 0.19823$ $a = 0.4$ $u = (z + 1/z)/2$ $v = u(1+p) + p$ $s =$ $-i\sqrt{v-1}\sqrt{v+1}$ $z_1 =$ $s + \sqrt{s-1}\sqrt{s+1}$ $t = \dfrac{z_1 + i\sin a}{\cos a}$ $w = (t + 1/t)/2$	$|a| < \pi/2$, $p > 0$, segment and arc of a cir- cumference	

3.3   Catalog 3. Curvilinear angular domains

In this catalog for each curvilinear angular domain the function $f(z)$ maps the half-plane $\operatorname{Im} z > 0$ onto the considered domain and satisfies the condition $f(\infty) = \infty$.

#	Function	Comment	Figure
1	$p = 1.5$ $$r = \frac{z-1}{z+1}$$ $s = r^p$ $$w = \frac{1+s}{1-s}$$	$0 < p < 2$, segment of a circle	
2	$h = 0.5$ $p = 1.6$ $t = i\sqrt{h^2 - z^2}$ $$r = \frac{t-1}{t+1}$$ $s = r^p$ $$w = \frac{1+s}{1-s}$$	$h > 0$, $0 < p < 2$, circular lune with a cut	
3	$p = 0.75$ $r = z + i\sqrt{1-z^2}$ $$s = \frac{r-i}{r+i}$$ $t = s^{2p}$ $$w = i\frac{1+t}{1-t}$$	$0 < p < 1$, two arcs of a circumference	
4	$p = 1.6952$ $r = z + i\sqrt{1-z^2}$ $$s = \frac{ip}{2}\left(r - \frac{1}{r}\right)$$ $$t = \frac{p}{p^2+1}\left(s + \frac{1}{s}\right)$$ $w = -it + \sqrt{1-t^2}\,\mathrm{sign}\,(\mathrm{Re}\,z)$	$p > 1$	

#	Function	Comment	Figure
5	$q = 0.7$ $p = -0.7$ $s = \dfrac{1}{2}\left(z + \dfrac{1}{z}\right)$ $t = \dfrac{s\,(q - p)}{2} + \dfrac{q + p}{2}$ $w = t + i\,\sqrt{1 - t^2}$	$-1 \le p <$ $q \le 1$, two arcs of a circumference	
6	$a = 0.7114$ $p = 1$ $s = i\,\sqrt{p^2 - z^2}$ $t = \dfrac{\cos a}{2}\left(s + \dfrac{1}{s}\right)$ $w = t + i\,\sqrt{1 - t^2}$	$p > 0$, $0 < a < \dfrac{\pi}{2}$, two arcs of a circumference and a cut	
7	$p = 1.5$ $a = \dfrac{\pi}{2}\dfrac{p - 1}{p + 1}$ $r = \sin a - \dfrac{1}{z}$ $s = \arcsin r$ $t = \dfrac{1 + p}{\pi}\,s + \dfrac{1 - p}{2}$ $w = -\dfrac{1}{t}$	$p > 0$, two tangent half-circles	
8	$p = 1.4$ $a = p - 0.5\sin(2p)$ $r = \cos p - \dfrac{1}{z}$ $s = $ $\arccos r - r\,\sqrt{1 - r^2}$ $w = \dfrac{a}{s - a}$	two tangent half-circles	

#	Function	Comment	Figure
9	$b = 0.5$ $a = 1$ $p = 0.5$ $t = i\sqrt{p^2 - z^2}$ $w = at + ib\sqrt{1 - t^2}$	$a, b, p \geq 0$, $a + b > 0$, ellipse with a cut	
10	$p = -0.4$ $t = \sqrt{z}$ $w = (t + p + i)^2$	arcs of confocal parabolas	
11	$r = 1 - \dfrac{1}{z}$ $s = \operatorname{arch} r$ $w = -\left(\dfrac{\pi}{s}\right)^2$	half of a cardioid	
12	$r = 1 - z^2$ $s = r + r\sqrt{1 - \dfrac{1}{r^2}}$ $w = i\sqrt{s + 1}$	loop of Bernoulli's lemniscata	

#	Function	Comment	Figure
13	$p = 1.02$ $u = z + i\sqrt{1 - z^2}$ $t = p\,u$ $w = t\sqrt{1 + \dfrac{1}{t^2}}$	$p = 1$ — half of Bernoulli's lemniscata $p > 1$ — half of single-contour cassinian	
14	$p = 0.95$ $h = \dfrac{1}{2}\left(p - \dfrac{1}{p}\right)$ $v = i\sqrt{h^2 - z^2}$ $u = v + i\sqrt{1 - v^2}$ $t = p\,u$ $w = t\sqrt{1 + \dfrac{1}{t^2}}$	$0 < p < 1$, half of two-contour cassinian	
15	$t = 1 - i z$ $w = i\left(t + \dfrac{1}{t} - 2\right)$	cissoid	
16	$r = \dfrac{i}{z\sqrt{2}}$ $s = \operatorname{arch} r$ $t = \left(\dfrac{2s}{\pi}\right)^2$ $w = \dfrac{4i}{t + 1} - i$	cissoid	

#	Function	Comment	Figure
17	$s = z + i\sqrt{1 - z^2}$ $t = i\,s$ $w = -i\,t\,\dfrac{t+1}{t-1}$	strophoid	
18	$q = 3$ $w = i\,q\sqrt{1 - z^2} +$ $\operatorname{arch} z - i\pi$	$q > 1$, step with a thin barrier	
19	$t = \left(1 + \dfrac{1}{z}\right)^{1/4}$ $w = 2t\,z + \operatorname{arth} t -$ $\arctan\dfrac{1}{t}$	slope step	
20	$t = \left(1 + \dfrac{1}{z}\right)^{1/4}$ $w = \dfrac{2t^3}{3(t^4 - 1)} +$ $\operatorname{arth} t - \arctan t$	hanging step	

#	Function	Comment	Figure
21	$c = 1.054$ $t = \dfrac{c}{z} + 2\operatorname{arth}\dfrac{1}{z}$ $w = \dfrac{2\pi}{t}$	half-circumference tangent to a straight line	
22	$c = 1.054$ $r = z + i\sqrt{1 - z^2}$ $s = \dfrac{i}{2}\left(r - \dfrac{1}{r}\right)$ $t = \dfrac{c}{s} + 2\operatorname{arth}\dfrac{1}{s}$ $w = -\dfrac{2i\pi}{t}$	cut along arcs of tangent circumferences	
23	$a = 0.75$ $t = \sqrt{\dfrac{z}{a^2 z + 1}}$ $u = \dfrac{1 + a\,t}{1 - a\,t}$ $w = u\left(\dfrac{t - 1}{t + 1}\right)^a$	$0 < a < 1$, sector bulge	
24	$a = 1.25$ $t = \sqrt{\dfrac{z}{a^2 z + 1}}$ $u = \dfrac{1 + a\,t}{1 - a\,t}$ $w = u\left(\dfrac{1 - t}{1 + t}\right)^a$	$1 < a < 2$, sector hollow	

#	Function	Comment	Figure
25	$m = 0.1$ $a = 1 - \dfrac{E'(m)}{K'(m)}$ $w =$ $E(z, m) - a\,F(z, m)$	$0 < m < 1$, two vertical segments	
26	$m = 0.04$ $u = F(z, m)$ $s = \dfrac{\pi}{K'(m)}\,u$ $w = \tanh s$	$0 < m < 1$, two half-circle bulges	
27	$a = 0.43$ $m = 0.81$ $w =$ $E(z, m) - a\,F(z, m)$		
28	$w = z + i\,e^{iz} - i$	cycloid	

#	Function	Comment	Figure
29	$m = 0.5$ $r = z + i\sqrt{1 - z^2}$ $s = \sqrt{r}$ $u = \dfrac{1}{2}\left(s + \dfrac{1}{s}\right)$ $h = \sqrt{2}\,(E(m) - m\,K(m))$ $t = E(u, 2) + i\,h\,\mathrm{sign}\,(\mathrm{Im}\,u)$ $w = t^2$	parabolic lens	
30	$m = 0.5$ $s = z + i\sqrt{1 - z^2}$ $r = \sqrt{i\,s}$ $u = \dfrac{1}{2}\left(r + \dfrac{1}{r}\right)$ $h = \dfrac{E(m) - m\,K(m)}{\sqrt{m}}$ $t = E(u, 2) + i\,h\,\mathrm{sign}\,(\mathrm{Im}\,u)$ $w = -i\,t^2$	segment of parabola	
31	$m = 0.5$ $s = 1 - z^2$ $\zeta = s + \sqrt{s - 1}\sqrt{s + 1}$ $r = \sqrt{i\zeta}$ $u = (r + 1/r)/2$ $h = \dfrac{E(m) - m\,K(m)}{\sqrt{m}}$ $t = E(u, 2) + i\,h\,\mathrm{sign}\,(\mathrm{Im}\,u)$ $w = i\sqrt{-i\,t^2}$	isosceles right-angle triangle	
32	$p = 0.5$ $t = p^2 - z^2$ $w = \left(1 + \sqrt{t}\right)^2$	$p \geq 0$, exterior of a parabola and a cut	

#	Function	Comment	Figure
33	$p = 1.15$ $t = p^2 - z^2$ $w = t + \sqrt{t-1}\sqrt{t+1}$	$p \geq 1$	
34	$a = 1.21$ $b = \sqrt{a^2 - 1}$ $s = a - z^2$ $t = \sqrt{s-b}\sqrt{s+b}$ $w = t + \sqrt{t-1}\sqrt{t+1}$	$a > 1$	
35	$s = \dfrac{1 - \sqrt{2}\,z^2}{1 + z^2}$ $t = \sqrt{s-1}\sqrt{s+1}$ $w = \dfrac{t-1}{t+1}$		
36	$c = 1.200033$ $h = \sqrt{c^2 - 1}$ $r = -z^2$ $s = \dfrac{cr + h}{r + 1}$ $t = \sqrt{s-h}\sqrt{s+h}$ $w = \dfrac{1+t}{1-t}$	$c > 1$	

#	Function	Comment	Figure
37	$c = \sqrt{2}$ $s = c\,\dfrac{1 - z^2}{1 + z^2}$ $t = \sqrt{s-1}\sqrt{s+1}$ $w = \dfrac{t-1}{t+1}$	trident	
38	$t = \ln\left(1 - \dfrac{1}{z^2}\right)$ $w = \dfrac{\pi}{t}$	two tangent circles and negative real half-axis	
39	$t = \dfrac{1}{z} - \ln\left(1 + \dfrac{1}{z}\right)$ $w = \dfrac{\pi}{t}$	circle and positive real half-axis	
40	$p = 0.75$ $u = 1 - z^2$ $r = u + \sqrt{u-1}\sqrt{u+1}$ $s = \dfrac{r-1}{r+1}$ $t = s^{2p}$ $w = \dfrac{1+t}{1-t}$	symmetrical circular lune	

#	Function	Comment	Figure
41	$s = \dfrac{z}{z+1}$ $t = \sqrt{s}\,(3 - s)$ $w = \dfrac{2}{t-2}$	half-circle and negative real half-axis	
42	$t = \left(1 + \dfrac{1}{z}\right)^{1/4}$ $u = \dfrac{2}{3}\,(4z+3)\,z\,t^3$ $w = u - \operatorname{arth}\dfrac{1}{t} -$ $\arctan\dfrac{1}{t}$		
43	$w = -i\,\arcsin z$	half-strip	
44	$p = 1$ $t = i\,\sqrt{p^2 - z^2}$ $w = -i\,\arcsin t$	$p > 0$ half-strip with a cut	

#	Function	Comment	Figure
45	$s = -1 - z^2$ $t = \text{arch } s$ $w = t^2$	parabola	
46	$t = \text{arch } z$ $w = t^2$	a half of parabola	
47	$p = 2$ $s = p^2 - 1 - z^2$ $t = \text{arch } s$ $w = t^2$	$p > 0$ parabola and a cut	
48	$p = \pi/2$ $s = -i \arcsin z$ $t = s + p$ $w = t^2$	$p \geq 0$	

#	Function	Comment	Figure
49	$t = \left(\dfrac{z}{z+1}\right)^{1/4}$ $w = \operatorname{arth} t + \arctan t$		
50	$s = -z^2$ $t = \left(\dfrac{s}{s+1}\right)^{1/4}$ $w = \operatorname{arth} t + \arctan t$		
51	$t = \left(1 - \dfrac{1}{z^2}\right)^{1/4}$ $w =$ $\arctan t + \operatorname{arth} \dfrac{1}{t} - \dfrac{\pi}{2}$		
52	$t = 1 - i z$ $w = \ln t - 1$	catenary of a constant resistance	

#	Function	Comment	Figure
53	$p = 0.75$ $t = \sqrt{1 - z^2} - i\,z$ $w = t^{2p}$	$0 < p < 1$, angle with a sector deleted	
54	$p = 0.85$ $s = -i\,z$ $t = 2p \arccos s$ $w = \cos t$	$0 < p < 1$, hyperbola	
55	$p = 0.75$ $t = p \arccos z$ $w = \cos t$	$0 < p < 1$, half of a hyperbola branch	
56	$p = 0.625$ $s = z^p$ $r = 1/\sin(\pi p)$ $t = s + re^{i\pi p}$ $w = t^2$	$0 < p < 1$, arc of the parabola and a ray of the straight line	

#	Function	Comment	Figure
57	$s = 1 - \dfrac{1}{z}$ $t = is\sqrt{1-s^2} - \text{arch } s$ $w = \dfrac{\pi}{t}$	right angle with half-circular cut-out	
58	$s = i\sqrt{z}$ $t = \dfrac{s}{z+1} + \text{arth } \dfrac{1}{s}$ $w = \dfrac{\pi}{2t}$	right angle with half-circular bulge	
59	$p = 1.8$ $w = z^p + (z-1)^p$	$1 < p < 2$, rounded angle	

3.4   Catalog 4. Curvilinear strips

In this catalog for each curvilinear strip the function $w = f(z)$ maps the strip $0 < \text{Im } z < \pi$ onto the considered domain and points $z = -\infty$ and $z = \infty$ — onto the branches of the strip. The function $f(z)$ is determined within the argument shift. Indeed, the function $w = f(z+c)$, where c is a real constant, carries out the same mapping.

#	Function	Comment	Figure
1	$a = -0.6$ $b = 0.8$ $s = \cosh z$ $t = \dfrac{1}{2}[s\,(b-a)+b+a]$ $w = i \arcsin t$	$-1 \leq a <$ $b \leq 1$, strip with two transverse cuts	
2	$p = 0.5$ $s = \cosh z$ $t = p\,(s-1)-1$ $w = t\,\sqrt{1 - 1/t^2}$	$p > 0$, two rays and perpendicular segment	
3	$\alpha = 0.8$ $s = 0.75\,i\,\cosh(z/2)$ $t = s + i\,\sqrt{1 - s^2}$ $w = t^{2\alpha}$	$0 < \alpha < 1$	
4	$s = \cos iz$ $t = (s+1)/2$ $u = \dfrac{t\,\sqrt{2}+1}{1-t}$ $v = \sqrt{u-1}\sqrt{u+1}$ $w = \dfrac{v-1}{v+1}$	two rays and half-circumference	

#	Function	Comment	Figure
5	$p = 0.5$ $u = \cos iz$ $r = p\,u$ $s = \dfrac{1}{2}\left(r + \dfrac{1}{r}\right)$ $t = \dfrac{2p\,s}{1+p^2}$ $w = t +$ $i\sqrt{1-t^2}\,\text{sign}\,(\text{Im}\,u)$	$0 < p < 1$, two symmetrical arcs of a circumference and two rays of the real axis	
6	$b = 2$ $p = 1.8$ $s = \cosh z$ $t = \dfrac{1}{2}[s\,(p-1)-p-1]$ $u = t + \sqrt{t-1}\sqrt{t+1}$ $w = \dfrac{(1+b)\,u}{2} + \dfrac{1-b}{2u}$	$p > 1$, $b > 0$, ellipse with half-axes 1 and b and two rays of the real axis	
7	$a = 1$ $b = 0.7$ $p = 0.9$ $t = p\,\sinh(z/2)$ $w = a\,t + i\,b\,\sqrt{1-t^2}$	$a, b > 0$, $p > 0$, elliptical bulge with half-axes a and b and a cut	
8	$q = 1$ $p = 0.75$ $u = \cosh z$ $v = q\,(1-u)$ $t = 2p\,\arccos\sqrt{v}$ $w = \cos t$	$q > 0$, $0 < p < 1$, branch of hyperbola and a cut	

#	Function	Comment	Figure
9	$a = 0.7$ $u = \cosh z$ $s = 0.5\,a\,(u+1)$ $t = 1 + \sqrt{s}$ $w = t^2$	$a > 0$, parabola and a cut	
10	$p = 10$ $u = \cosh z$ $s = \dfrac{1}{2}[u\,(p+1)+p-1]$ $t = \arccos s$ $w = t^2$	$p > -1$, parabola and a cut	
11	$s = e^z$ $t = 1 + s^2 - \ln(-1/s^2)$ $w = i\,\sqrt{t}\,\mathrm{sign}\,(\mathrm{Re}\ s)$	two branches of non-confocal hyperbolas	
12	$r = e^z$ $s = \dfrac{1+r}{1-r}$ $u = \dfrac{1}{2}\left(s + \dfrac{1}{s}\right)$ $v = \sqrt{u}$ $t = \mathrm{arth}\,v + \arctan v$ $w = t^2$	four arcs of confocal parabolas	

#	Function	Comment	Figure
13	$t = \dfrac{2z}{3\pi} - \dfrac{i}{3}$ $w = t^3 + \dfrac{t}{3}$	two Neil's parabolas	
14	$a = 0.9$ $p = 0.25\,(a + 1/a - 2)$ $u = \cosh z$ $r = p\,u - p - 1$ $s = r + \sqrt{r+1}\sqrt{r-1}$ $t = 1 + a\,s$ $w = \sqrt{t}\,\mathrm{sign}\,(\mathrm{Im}\,z)$	$0 < a < 1$, two-contour cassinian and two rays	
15	$t = z - i\,\dfrac{\pi}{2}$ $w = t + e^t$	Rogowskii's capacitor (two exponential curves)	
16	$t = z - i\,\dfrac{\pi}{2}$ $w = t + \sinh t$	two catenaries	

#	Function	Comment	Figure
17	$w = z - \tanh(z/2)$	two tractrices	
18	$p = 5$ $a = 0.667$ $b = 0.667$ $t = e^z$ $w =$ $(1 - 1/t)^a \, (t - p)^b$	$p > 1$, $a, b > 0$, $a + b \leq 2$, star-shaped cut	
19	$a = 2$ $w = z + a \cosh z - i\dfrac{\pi}{2}$	$a > 0$, two rays of parallel straight lines	
20	$b = 1$ $t = e^z$ $w = -t^2 + 2\ln t + b\,t +$ $1 - i\,\pi$	two rays of parallel straight lines	

#	Function	Comment	Figure
21	$a = 0.75$ $p = 1.3$ $t = e^z$ $s =$ $\dfrac{t^2}{2-a} + \dfrac{t(p-1)}{1-a} + \dfrac{p}{a}$ $w = \dfrac{s}{t^a}$	$0 < a < 2$, $p > 0$, two cuts along rays of intersecting straight lines	
22	$p = 2.6$ $q = 1.4$ $z_1 = e^z$ $t = \sqrt{\dfrac{p^2 - z_1}{z_1 - 1}}$ $s =$ $\arctan t + q \arctan \dfrac{t}{p}$ $w = i\left(s - \dfrac{\pi}{2}\right)$	$p, q > 1$, half-strip with a cut	
23	$p = 1$ $r = \cosh z$ $s = p(1 - r)$ $t = i\sqrt{s}$ $w = it\sqrt{1 - t^2} -$ $\operatorname{arch} t + i\pi/2$	exterior of half-strip with a cut	
24	$t = \sqrt{e^z + 1}$ $w = 2t + \ln\dfrac{1-t}{1+t}$		

#	Function	Comment	Figure
25	$t = (e^z + 1)^{1/4}$ $w = 4t + \ln\dfrac{1-t}{1+t} - 2\arctan t$		
26	$t = (e^z + 1)^{1/4}$ $w = \dfrac{4}{3}t^3 + \ln\dfrac{1-t}{1+t} + 2\arctan t$		
27	$p = 1$ $s = e^z$ $t = \sqrt{s+1}$ $w = t/p + 2\operatorname{arth} t$	$p > 0$	
28	$p = 0.2$ $s = e^z$ $t = \sqrt{s+1}$ $w = 2\operatorname{arth} t - t/p$	$0 < p < \dfrac{1}{2}$	

#	Function	Comment	Figure
29	$p = 0.4$ $s = e^z$ $r = \sqrt{\dfrac{s-1}{s+p^2}}$ $t = ipr$ $u = 2\,\mathrm{arth}\,r$ $v = i\,(1/p - p)\,\mathrm{arth}\,t$ $w = u + v - r\left(1 + \dfrac{p^2}{s}\right)$	$p > 0$	
30	$p = -1.4$ $s = e^z$ $r = i\sqrt{s-1}$ $w = r\left(\dfrac{p}{s} + 1\right) -$ $(p+1)\,\mathrm{arth}\,r$		
31	$p = 0.65$ $s = e^z$ $t = \sqrt{s/p + 1}$ $w = 2t\,(s + p - 3)/3 +$ $2\,\mathrm{arth}\,(1/t)$	$p > 0$	
32	$p = 1.5$ $s = e^z$ $t = \sqrt{\dfrac{s-1}{s+p^2}}$ $w =$ $p\,\mathrm{arth}\,t - \arctan pt$	$p > 0$	

#	Function	Comment	Figure
33	$p = 0.5$ $s = e^z$ $t = \sqrt{\dfrac{s-1}{s-p^2}}$ $w =$ $p \arctan t - \operatorname{arth} p\,t$	$0 < p < 1$	
34	$p = 0.5$ $s = e^z$ $t = \sqrt{\dfrac{s+p^2}{s+1}}$ $u = \left(p + \dfrac{1}{p}\right) \arctan t$ $v = \left(\dfrac{1}{p} - p\right) \dfrac{t}{1-t^2}$ $w = -2 \operatorname{arth}(t/p) +$ $u + v$	$0 < p < 1$	
35	$p = 2$ $u = \cosh z$ $v = p(u+1)$ $r = i\sqrt{v}$ $t = \left(1 - \dfrac{1}{r^2}\right)^{1/4}$ $w = \operatorname{arth} \dfrac{1}{t} - \arctan t$	$p > 0$	
36	$m = 0.4$ $u = \cosh z$ $w = E(u, m) - E(m)$	$0 < m < 1$	

#	Function	Comment	Figure
37	$m = 0.1$ $t = \cosh z$ $a = 1 - \dfrac{E'(m)}{K'(m)}$ $w =$ $E(t, m) - a\, F(t, m)$	$0 < m < 1$	
38	$m = 0.9$ $a = -1$ $t = \cosh z$ $w =$ $E(t, m) - a\, F(t, m)$	$0 < m < 1$	
39	$m = 0.5$ $s = e^z$ $t = \sqrt{s + 1}$ $u = t \sqrt{m - \dfrac{1 - m}{s}}$ $w = u +$ $F(t, m) - 2E(t, m)$	$0 < m < 1$	
40	$m = 0.02$ $u = \cosh z$ $r = F(u, m)$ $s = \dfrac{\pi}{K'(m)}\, r$ $t = e^s$ $w = \dfrac{1 - t}{1 + t}$	$0 < m < 1$	

3.5 Catalog 5. Curvilinear strips with N branches

In this catalog the mappings of curvilinear strips with three (domains 1–6) and four branches (domains 7–14) are presented. For the domains 1–5 the corresponding function maps the half-plane $\text{Im } z > 0$ onto the considered domain. In the comment the points of the real axis a_1, a_2, a_3 are shown. These points are the preimages of branches of the curvilinear strip. They are necessary for the construction of flows without branching (Section 2.13) and flows with branching (Section 2.20).

For the domains 6–12 the function which maps the half-plane $\text{Im } s > 0$ onto the considered domain is also presented in the catalog. In the comment the points of the real axes s_k are preimages of the branches of the strip. For the domains 6–12 that do not have straight angles the mapping of the half-plane onto the considered domain is not visual and does not have physical sense. For these domains the mappings of the strip $0 < \text{Im } z < \pi$ are presented that map the points $z = -\infty$ and $z = \infty$ into two branches of the strip (flows without branching). The variable s belonging to the half-plane $\text{Im } s > 0$ is considered as auxiliary variable. It is possible to construct other flows without branching in the considered strips with the help of the method stated in Section 2.13.

At last the mappings of the circle $|z| < 1$ onto the symmetrical domains 13–14 are presented in the catalog. The comment shows the points of the unit circumference z_k that are the preimages of the branches of the strip.

#	Function	Comment	Figure
1	$p = 0.2785$ $q = 0$ $h = 1$ $t =$ $h \ln(z+1) + \ln(1-z)$ $w = t - p z^2 + q z$	$p, h > 0$ $a_1 = -1$ $a_2 = 1$ $a_3 = \infty$	
2	$p = 1$ $q = 1$ $h = 1.5$ $u = h \ln(z+1) - \ln z$ $w = u + p/z + q z$	$p, q, h > 0$ $a_1 = -1$ $a_2 = 0$ $a_3 = \infty$	
3	$c = 1$ $h = 1.4$ $w = c z + h \ln(z+1) -$ $\ln(z-1)$	$c, h > 0$ $a_1 = -1$ $a_2 = 1$ $a_3 = \infty$	
4	$c = 2$ $h = 0.6$ $w = c z + h \ln z +$ $\ln(1-z)$	$c, h > 0$ $a_1 = 0$ $a_2 = 1$ $a_3 = \infty$	

#	Function	Comment	Figure
5	$a = 0.3$ $b = 0.4$ $c = 1.5$ $p = 2$ $w = \dfrac{z^c}{(z+p)^a \, (z+1)^b}$	$0 < c < 2$, $a, b > 0$, $a + b < c$, $p > 1$, star-shaped domain	
6	$h = 0.8$ $s = \tanh(z/2)$ $w = \ln(s+1) +$ $h \ln(s - t)$	$h > 0$ $s_1 = -1$ $s_2 = 1$ $s_3 = \infty$	
7	$p = 1$ $s = \tanh(z/2)$ $t = \arcsin s$ $w = t + p \tan t$	$p > 0$ $s_1 = -1$ $s_2 = 1$ $s_3 = \infty$	
8	$c = 1.4142$ $s = \tanh(z/2)$ $r = \sqrt{s+c}\,\sqrt{s-c}$ $t = s/r$ $w = i\,(\arctan t +$ $\operatorname{arth} t)$	$c > 1$ $s_1 = -c$ $s_2 = c$ $s_3 = \infty$	

#	Function	Comment	Figure
9	$p = 1.3$ $c = \dfrac{p}{\sqrt{p^2 - 1}}$ $s = p \tanh(z/2)$ $t = \sqrt{1 - \dfrac{1}{s^2}}$ $w = $ $c \operatorname{arth} \dfrac{1}{t} - \operatorname{arth} \dfrac{1}{ct}$	$p > 1$ $s_1 = -p$ $s_2 = p$ $s_3 = \infty$	
10	$p = 1$ $h = 0.7$ $s = e^z$ $w = \ln(s+1) + h \ln s - \ln(s - p)$	$p, h > 0$ $s_1 = -1$ $s_2 = 0$ $s_3 = 1$ $s_4 = \infty$	
11	$a = 2$ $h = 0.3$ $s = e^z$ $u = \dfrac{a}{s^2 - 1} + \ln s$ $v = h\left[\ln(1 - s) + \ln(1 + s)\right]$ $w = u + v - i\pi/2$	$a, h > 0$ $s_1 = -1$ $s_2 = 0$ $s_3 = 1$ $s_4 = \infty$	
12	$s = \tanh(z/2)$ $w = \dfrac{s}{\sqrt{1 - s^4}}$	mapping of the strip $0 < \operatorname{Im} z < \pi$	

#	Function	Comment	Figure
13	$t = z\sqrt{\dfrac{2}{1+z^4}}$ $w = \arctan t + \operatorname{arth} t$	mapping of the circle $\lvert z \rvert < 1$ $z_1 = -1$ $z_2 = -i$ $z_3 = 1$ $z_4 = i$	
14	$r = i\,z^2$ $s = -i\arctan r$ $w = \sqrt{s}\,\operatorname{sign}(\operatorname{Re} z)$	mapping of the circle $\lvert z \rvert < 1$ $z_1 = -1$ $z_2 = -i$ $z_3 = 1$ $z_4 = i$	

A

Appendix: Program CONFORM

The program CONFORM is supplied in the floppy disk enclosed with the book. It is necessary to install the CONFORM program in the hard disk. About 2Mb of free disk space is necessary for the installation.

Other hardware and software requirements are as follows:

1. IBM PC computer with 286, 386, 486 or Pentium processor or fully compatible,

2. 4 Mb of free hard disk space is recommended to improve performance,

3. the presence of a coprocessor is desirable to make the plotting of a mapping faster (note that most of the 486 family of processors and the Pentium processor have built-in coprocessors),

4. graphics card supported by MS Windows,

5. Microsoft Windows version 3.1 or higher installed (the CONFORM will run under MS Windows version 3.0, but some messages will not be correct because of absence of True Type Fonts).

6. any printer supported by MS Windows if you are going to print obtained figures.

Installation

Insert the diskette with the program CONFORM into the floppy disk drive and type from the MS DOS prompt (do not load MS Windows before):

    A:\INSTALL

(if the current floppy disk drive is the drive A:) or

    B:\INSTALL

(if this drive is B:).

INSTALL program will check the software and hardware environment and it can fail in the case when there is not enough free memory, not enough hard disk space, MS Windows is not installed at the computer or you have started installation running the MS DOS prompt from MS Windows. In this case solve the problem and start installation again. If installation is possible you will be prompted to enter the desired path where you want CONFORM program to be installed. Enter it or press the **Enter** key if the default path C:\CONFORM is suitable for you. Please wait for several minutes for all necessary files to be copied into the hard disk. After finishing the installation MS Windows will be started automatically and a new group with the CONFORM icon (a small cobweb) will be created.

How to use the program CONFORM

We suppose that the reader is familiar with MS Windows 3.1. If it is not true, please refer to your MS Windows manual. All MS Windows programs use the similar interface, in other words, all programs are controlled by the similar mouse and keyboard commands. It makes it easy to learn the new program, thus we will describe only specific features of the program CONFORM. It has three modes which are called **Mapping**, **Isoline** and **Flow** modes. Use **Mode** item of the main menu to switch between modes.

In all modes the main loop of the work consists of the entering and editing of function(s) using **Edit** item of the main menu (F3 hot key), compiling it using **Compile** command (F9 hot key), adjusting parameters (**Parameters** subitem of menu) and plotting of the image (**Run** command or F10 hot key).

Menu commands overview

Files and print. Similarly to other MS Windows programs it is possible to save your work in a file and to open previously saved files in all three modes. These files have the extension **MAP** for the **Mapping** mode, **ISO** for **Isoline** mode and **STR** for the **Flow** mode. If you save a file without extension, an appropriate extension will be added in correspondence with currently selected mode. **New** command opens a new work clearing all the editors and assigning default values to all the parameters. If previous work was not saved you will be prompted to do this. The same is true for the **Exit** command.

Sometimes it is useful to print a result. Check whether the installed

MS Windows printer driver corresponds to the currently available printer. Use **Printer setup** command for this purpose. Also check whether the size of the paper is chosen accordingly. To obtain high quality output it is recommended to use the highest resolution of the printer. You can also vary left, right, top and bottom margins of the output. Use **Page setup** command for the purpose. Note that the output always resides in square frame unlike the CONFORM window at the screen. Printing of an image on a laser printer with high resolution can take up to several minutes, so be patient, please.

Edit menu. **Function** item (hot key F3) lets you open function editor and enter or modify the function. There are two editors for the **Isoline** mode, the first (upper) is for the potential and the second is for the mapping function. In this mode **Function** command (F3 hot key) also toggles between these editors.

Compile command (hot key F9) prepares the function for plotting. If syntax is correct, a message "Function compiled successfully" will appear at the bottom of CONFORM window. If something is wrong, an error message will appear and the cursor will be placed at the place where the error is likely to be present.

Catalog command lets you open one of five catalogs containing useful mappings. Note that only the second catalog is accessible in the **Flow** mode.

Other commands are common for Windows programs and let you cut, copy and paste a part of the text. Using clipboard it is possible to transfer the functions between CONFORM and other Windows applications (Notepad, for example). It is a way to import or export functions from or to text files. Therefore, it is possible to use a mapping function as a part of another program (written in FORTRAN, for example).

Syntax of function

CONFORM program uses FORTRAN-like syntax for functions. The function should be written as a sequence of assignment operators, each at the separate string. Each operator defines a new variable placed in the left-hand side of the assignment. The last defined variable is used as the output value and is plotted at the screen. Do not use predefined names Z, I, E and K at left-hand side of the expression. The right-hand side of the assignment operator can consist of already defined variables combined with the help of arithmetic operators $+$, $-$, $*$, $/$, power operator $**$ and built-in functions of a complex variable. Precedence of evaluation is as follows:

1. all built-in functions are evaluated;

2. all power operations in the order from left to right are evaluated. Thus, A**B**C means $(A^B)^C$, not $A^{(B^C)}$;

3. all multiplications (∗) and divisions (/) are evaluated from left to right;

4. all additions (+) and subtractions (−) are evaluated from left to right.

At any time these rules can be altered using the parentheses. For example, linear-fractional function $F = \frac{1+Z}{1-Z}$ can be written as F=(1+Z)/(1-Z).

By convention the initial variable is designated as Z. The imaginary unit I is the predefined variable, thus complex constant $a = 2.1 + 0.2i$ can be written as A=2.1+0.2*I, for example.

The number of strings in the expression is not limited. The length of the separate string is not limited also. Nevertheless it is recommended to split complicated expressions into several more simple ones. See functions in catalogs of the CONFORM program as examples of how to do this. You can compare functions defined in catalogs with the functions printed in the book. Simply load the file and try to experiment with it.

The following built-in functions of one complex variable can be used:

REAL(Z) — the real part of a complex number Re z;

IMAG(Z) or AIMAG(Z) — the imaginary part of a complex number Im z;

ABS(Z) or CABS(Z) — modulus of a complex number $|z|$;

ARG(Z) — the argument of a complex number arg z;

CONJ(Z) — the complex conjugate of a complex number $\bar{z}$;

SIGN(Z) — the sign of the real part of a complex number $F = $ sign (Re z), i.e., $F = 1$ if Re $z > 0$, $F = -1$ if Re $z < 0$ and $F = 0$ if Re $z = 0$;

SQRT(Z) or CSQRT(Z) — $\sqrt{z}$;

EXP(Z) or CEXP(Z) — e^z;

LOG(Z) or CLOG(Z) — ln z;

LOG10(Z) or CLOG10(Z) — log z;

COS(Z) or CCOS(Z) — cos z;

SIN(Z) or CSIN(Z) — sin z;

TAN(Z) or CTAN(Z) — tan z;

COSH(Z) or CCOSH(Z) — cosh z;

SINH(Z) or CSINH(Z) — sinh z;

TANH(Z) or CTANH(Z) — tanh z;

ASIN(Z) or CASIN(Z) — arcsin z;

ACOS(Z) or CACOS(Z) — arccos z;

ATAN(Z) or CATAN(Z) — arctan z;

ASINH(Z) or CASINH(Z) — the hyperbolic arccosine arsh z;

ACOSH(Z) or CACOSH(Z) — the hyperbolic arcsine arch z;

ATANH(Z) or CATANH(Z) — the hyperbolic arctangent arth z;

K(Z) — the complete elliptic integral of the first kind of the real part of a complex number z: $K(\mathrm{Re}\ z)$ (Subsection 1.13.1);

E(Z) — the complete elliptic integral of the second kind of the real part of a complex number z: $E(\mathrm{Re}\ z)$ (Subsection 1.13.2).

The following built-in functions of two complex arguments can be used (only the real part of the second argument is taken into account):

ELLF(Z,M) — the elliptic integral of the first kind $F(z, m)$ (Subsection 1.13.1);

ELLE(Z,M) — the elliptic integral of the second kind $E(z, m)$ (Subsection 1.13.2);

SN(Z,M) — the elliptic sine sn (z, m) (Subsection 1.13.5);

CN(Z,M) — the elliptic cosine cn (z, m) (Subsection 1.13.5);

DN(Z,M) — the delta amplitude dn (z, m) (Subsection 1.13.5).

Note that the principal branch of the multiply-valued functions is calculated by the program CONFORM.

Catalogs

A number of ready-to-use mapping functions together with concrete values of parameters are stored in catalogs. Catalogs are numbered in the correspondence with the Chapter 3. Make **Catalog** command and select the desired Catalog from the list. You will see the Catalog dialog box. It is possible to browse through the catalog. If **Auto Redraw** check box is checked, each new selection in the list of available files will be shown at the screen as the image of the Cartesian or the polar net. For the first time the browsing is rather slow, but CONFORM creates special files containing cached binary data. Therefore all subsequent browses through the catalog will be much faster. You can also use **Redraw** button for this purpose. It is possible to delete any file from the Catalog or to add a new one. Any file can be loaded into the program CONFORM for the experiments with the mapping using **Load** button.

It is recommended to start with loading of these files and with experiments with them.

Parameters

Menu command **Parameters** opens the Parameters dialog box. It looks slightly different in three different modes of the CONFORM.

In all the three modes the parameters of the net are to be entered. They include the type of the net (Cartesian or polar), the minimum and maximum values of coordinates and the number of nodes.

For the better understanding try to do a simple exercise: select **Mapping** mode, select **New** file, enter the simplest expression F=Z into the function editor, compile it (press F9) and plot the corresponding net (press F10). Try to vary the parameters describing the net and the type of the net.

Scale factor (or simply **Scale**) is used to choose the appropriate size of the image at the screen. The image can be scrolled with the help of scroll bars and the mouse. You can magnify the image three times by pressing the left mouse button inside the CONFORM window. The origin of a new plot will be shifted to the point where mouse cursor is located. The initial settings can be restored with the help of the right mouse button.

X and Y reflections can be used to see the mapping of several parts of the net simultaneously. It is especially useful when the function has branch points or cuts of the domain of the definition. Check whether the domains do not overlap under the reflection, in the opposite case you will see strange results at the screen.

In the **Isoline** mode a new selection is added. The user can select what should be plotted: the real or the imaginary part of the function (defined in the first editor) or both of the characteristics. The number of level lines can be entered also (the default value is 30).

For example, to plot the isothermic net of $\sin z$ you should enter into the first editor F=SIN(Z), then press F3 key to switch to the second (lower) editor and enter the identity transformation F=Z into it. Note that these editors are quite independent, thus the same variable names do not cause any ambiguity. Compile the expressions and plot the result. Try to experiment with all parameters. To plot the relief map of the same function add the second string to the first editor. It will look like:

$$F = \text{SIN}(Z)$$

$$W = \text{ABS}(F) + I * \text{ARG}(F)$$

Try to experiment with parameters again. The **Flow** mode is a particular case of **Isoline** mode with the predefined expression of the potential Ψ (2.18.5) (the first editor of the mode **Isoline**). Therefore, the circulation and the attack angle are added to the dialog box.

Options

Options commands shows the Options dialog box. The style and color of the axis and the lines of the net can be selected.

Where to get more information

Many figures of this book were obtained with the help of the program CON-FORM. Interesting files are collected in subdirectories FIGS1 and FIGS2 on your diskette. To save the disk space these directories are not installed in the hard disk, so copy them manually or load files from the floppy disk directly if you intend to use them.

Bibliography

[1] Schubert, F. T., De projectione sphaeroidis ellipticae geographica, *Nova Acta Academiae Scientiarum Imperialis Petropolitanae*, 5, 130, 1789.

[2] Gauss, C. F., Untersuchungen ueber Gegenstaende der hoehern Geodaesie (1844), *Werke*, Bd.4., Gesellschaft der Wissenschaften, Goettingen, 1880.

[3] Gauss, C. F., Allgemeine Aufloesung der Aufgabe: die Theile einer gegebenen Flaeche so abzubilden, dass die Abbildung dem Abgebildeten in den kleinsten Theilen aehnlich wird (1822), *Werke*, Bd. 4, Gesellschaft der Wissenschaften, Goettingen, 1880.

[4] Jacobi, C. G. J., Ueber die Abbildung eines ungleichaxigen Ellipsoids auf einer Ebene, bei welcher die kleinsten Theile aehnlich bleiben, *Gesammelte Werke*, Bd. 2, Reimer, Berlin, 1882, 399.

[5] Schwarz, H. A., Conforme Abbildung der Oberflaeche eines Tetraeders auf die Oberflaeche einer Kugel (1868), *Gesammelte mathematische Abhandlungen*, Bd. 2, Springer, Berlin, 1890, 84.

[6] Euler, L., Principes generaux du mouvement des fluides (1755), *Opera omnia*, Ser. II, V. 12, Teubner, Berlin-Leipzig, 1954.

[7] Euler, L., Ulterior disquisitio de formulis integralibus imaginariis (1777), *Opera omnia*, Ser. I, V. 19, Lausannae, 1932.

[8] D'Alembert, J. B., *Essai d'une nouvelle theorie de la resistance des fluides*, Paris, 1752.

[9] Euler, L., *Introduction to analysis of the infinite (1748)*, Springer, New York, 1990.

[10] Euler, L., De projectione geographica superficiei sphaericae (1777), *Opera omnia*, Ser. I, V. 28, Teubner, Berlin-Leipzig, 1955.

[11] Lagrange, J. L., Sur la construction des cartes geographiques (1779), *Oeuvres*, T. 4., Gauthier-Villars, Paris, 1869.

[12] Weierstrass, C., Definition analytischer Funktionen einer Veraenderlichen vermittelst algebraischer Differentialgleichung (1842), *Mathematische Werke*, Bd. 1., Mayer and Mueller, Berlin, 1895.

[13] Riemann, B., Grundlagen fuer eine Allgemeine Theorie der Funktionen einer veraenderlichen complexen Groesse (1851). *Collected works*,
Dover, New York, 1953.

[14] Weierstrass, C., Ueber das sogenannte Dirichlet'sche Prinzip (1869),
Mathematische Werke, Bd. 2., Mayer and Mueller, Berlin, 1895.

[15] Hilbert, D., Das Dirichletsche Prinzip (1901), *Gesammelte Abhandlungen*, Bd. 3., Springer, Berlin, 1935.

[16] Schwarz, H. A., Zur Theorie der Abbildung (1869). *Gesammelte
mathematische Abhandlungen*, Bd. 2, Springer, Berlin, 1890, 108.

[17] Caratheodory, C., Elementarer Beweis fuer den Fundamentalsatz der
konformen Abbildungen (1914). *Gesammelte mathematische Scriften*,
Bd. 3, Beck, Muenchen, 1955, 273.

[18] Monge, G., Application d' analyse a la geometrie, *annotee par J.
Liouville*, ed. 5, suppl. 6. Bachelier, Paris, 1850.

[19] Hilbert, D., Zur Theorie der konformen Abbildung (1909). *Gesammelte Abhandlungen*, Bd. 3, Springer, Berlin, 1935.

[20] Christoffel, E. B., Sul problema delle temperature stationarie (1867).
Sopra un problema proposito da Dirichlet (1871). *Gesammelte mathematische Abhandlungen*, Bd. 1–2., Teubner, Leipzig, 1910.

[21] Schwarz, H. A., Ueber einige Abbildungsaufgaben (1869). *Gesammelte mathematische Abhandlungen*, Bd. 2., Springer, Berlin, 1890.

[22] Schwarz, H. A., Notizia sulla rappresentazione conforme di un' area
ellitica sopra un'area circolare (1869). *Gesammelte mathematische
Abhandlungen*, Bd. 2, Springer, Berlin, 1890, 102.

[23] Schwarz, H. A., Ueber diejenigen Faelle, in welchen die Gaussische
hypergeometrische Reihe eine algebraische Funktion ihres vierten El-
ementes darstellt (1872). *Gesammelte mathematische Abhandlungen*,
Bd. 2, Springer, Berlin, 1890, 211.

[24] Jacobi, C. G. J., Fundamenta nova theoriae functionum ellipticarum
(1829). *Gesammelte Werke*, Bd. 1, Reimer, Berlin, 1881, 49.

[25] Riemann, B., *Elliptische Funktionen*. Vorlesungen (1856). Teubner,
Leipzig, 1899.

[26] Weierstrass, C., Vorlesungen ueber die Theorie der elliptischen Funktionen. *Mathematische Werke*, Bd. 5., Mayer und Mueller, Berlin,
1895.

[27] Schwarz, H. A., *Formeln und Lehrsaetze zum Gebrauch der elliptischen Funktionen*, Springer, Berlin, 1893.

[28] Klein, F., *Vorlesungen ueber die Theorie der elliptischen Modulfunktionen*, Bd. 1–2., Teubner, Leipzig, 1890–1892.

[29] Fricke, R., Klein, F., *Vorlesungen ueber die Theorie der automorphen Funktionen*, Bd. 1–2., Teubner, Leipzig, 1926.

[30] Klein, F., *Vorlesungen ueber die Entwicklung der Mathematik im 19 Jahrhundert*, Bd. 1, Springer, Berlin, 1927.

[31] Poincare, H., Sur les fonctions fuchsiennes (1881–1882). *Oeuvres*, Vol.2, Gauthier-Villars, Paris, 1952.

[32] Chebyshev, P. L., On geographic maps construction (1856), *Full collection of works*, V. 5., Academy of Sci. Publ., Moscow, 1951, (in Russian).

[33] Grave, D. A., *On basic problems of mathematical theory of geographic maps construction*, Academy of Sci. Publ., Saint-Petersbourg, 1896, (in Russian).

[34] Riemann, B., Weber, H., *Die partielle Differential-Gleichungen der mathematischen Physik*, Bd. 1–2, Vieweg, Braunschweig, 1901.

[35] Helmholtz, H., Ueber discontinuirliche Fluessigkeitsbewegungen (1868). *Wissenschaftliche Abhandlungen*, Bd. 1, Barth, Leipzig, 1882, 146.

[36] Kirchhoff, G. R., Zur Theorie freier Fluessigkeitsstrahlen (1869). *Gesammelte Abhandlungen*, Barth, Leipzig, 1882.

[37] Kirchhoff, G. R., *Vorlesungen ueber mathematische Physik*, Bd. 3, Teubner, Leipzig, 1891.

[38] Kirchhoff, G. R., Ueber die stationaeren electrischen Stroemungen in einer gekruemmten leitenden Flaeche (1875), *Gesammelte Abhandlungen*, Barth, Leipzig, 1882, 56.

[39] Zhukovskii, N. E., Ueber die Konturen des Tragflaechen der Drachenflieger (1910–1912), *Collected papers*, V. 5, Gl. red. av. lit., Moscow, 1937.

[40] Zhukovskii, N. E., Theoretical foundation of aeronautics (1911), *Collected papers*, V. 6. Gostekhizdat, Moscow, 1950, (in Russian).

[41] Chaplygin, S. A., On a pressure of planeparallel stream on obstacle bodies (to the theory of airplane) (1910), *Collected papers*, V. 2, Gostekhizdat, Moscow, 1948, (in Russian).

[42] Zhukovskii, N. E., Kirchhoff's method modification for a determination of liquid motion in two dimensions with constant velocity (1890). *Collected papers*, V. 2. Gostekhizdat, Moscow, 1950, (in Russian).

[43] Chaplygin, S. A., A theory of grid wing (1914), *Collected papers*, V. 2. Gostekhizdat, Moscow, 1933, (in Russian).

[44] Zhukovskii, N. E., Vorticity theory of the propeller (1915) (third article), *Collected papers*, V. 4. Gostekhizdat, Moscow, 1950, (in Russian).

[45] Kochin, N. E., *Hydrodynamic theory of grids*, Gostekhizdat, Moscow, 1949, (in Russian).

[46] Kolosov, G. V., On some application of functions of a complex variable to plane problem of mathematical elasticity theory. Mattison, Yuriev, 1909, (in Russian).

[47] Muskhelishvili, N. I., Sulla deformazione plana di un cilindro elastico isotropo, *R.C. Accad. Lincei*, 31, N 12, 1922, 548.

[48] Muskhelishvili, N. I., *Some Basic Problems of the Mathematical Theory of Elasticity*, Acad. Sci. Publ., Leningrad, 1933, (in Russian).

[49] Pavlovskii, N. N., The theory of ground water motion under hydrotechnical constructions and its main applications (1922). *Collected works*, V. 2, Academy of Sci. Publ., Moscow, 1956, (in Russian).

[50] Lavrent'ev, M. A., Keldysh, M. V., General problem of rigid impact on water, Collection of papers on the topic of water surface impact. Works of TSAGI, ed. 152, Moscow, 1935, (in Russian).

[51] Sedov, L. I., On a rigid body impact, floating on the surface of a noncompressible liquid, *Works of TSAGI*, ed. 187, 1934, (in Russian).

[52] Sedov, L. I., *The theory of plane motions of ideal liquid*, Oborongiz, Moscow, 1939, (in Russian).

[53] Lavrent'ev, M. A., To the theory of conformal mappings (1934), *Selected works*, Mathematics and mechanics, Nauka, Moscow, 1990, (in Russian).

[54] Kantorovich, L. V., Krylov, V. I., *Approximate Methods of Higher Analysis*, P. Noordhoff, Groningen (Netherlands), 1958.

[55] Fil'chakov, P. F., *Approximate Methods of Conformal Mappings*, Naukova dumka, Kiev, 1964, (in Russian).

[56] Privalov, I. I., *Introduction to the Theory of Functions of a Complex Variable*. Nauka, Moscow, 1984, (in Russian) or
Priwalow I. I. *Einfuehrung in die Funktionentheorie*, Teubner, Leipzig, 1970, (in German).

[57] Markushevich, A. I., *Theory of Functions of a Complex Variable*, V. 1–3, Prentice-Hall, New York, 1965–1967.

[58] Shabat, B. V., *An Introduction to the Complex Analysis*, Nauka, Moscow, 1985, (in Russian) or
Chabat, B. V., *Introduction a l'analyse complexe*, T. 1–2, Mir, Moscou, 1990, (in French).

[59] Forsyth, A. R., *Theory of Functions of a Complex Variable*, Cambridge Univ. Press, 1918.

[60] Copson, E. T., *An Introduction to the Theory of Function of a Complex Variable*, Oxford University Press, London, 1946.

[61] Caratheodory, C., *Theory of Functions of a Complex Variable*, V. 1–2, Chelsea Publ., New York, 1954.

[62] Ahlfors, L. V., *Complex Analysis*, McGraw-Hill, New York, 1966.

[63] Caratheodory, C., *Conformal Representation*, Cambridge Univ. Press, London, 1952.

[64] Bieberbach, L., *Conformal Mapping*, Chelsea Publ., New York, 1953.

[65] Lewent, L., *Konforme Abbildung*, Teubner, Berlin, 1912.

[66] Julia, G., *Lecons sur la representation conforme des aires simplement connexes*, Gauthier-Villars, Paris, 1931.

[67] Julia, G., *Lecons sur la representation conforme des aires multiplement connexes*, Gauthier-Villars, Paris, 1934.

[68] Hurwitz, A., *Vorlesungen ueber allgemeine Funktionentheorie und elliptische Funktionen*, Springer, Berlin, 1964.

[69] Courant, R., *Dirichlet's Principle, Conformal Mapping and Minimal Surfaces*, Interscience publ., New York, 1950.

[70] Goluzin, G. M., *Geometrical Theory of Function of a Complex Variable*, Nauka, Moscow, 1966, (in Russian).

[71] Jenkins, J. A., *Univalent Functions and Conformal Mapping*, Springer, Berlin, 1958.

[72] Nehari, Z., *Conformal Mapping*, McGraw-Hill, New York, 1952.

[73] Lavrentiev, M. A., Shabat, B. V., *Methods of Complex Function Theory*. Nauka, Moscow, 1987, (in Russian) or
Lavrentiev, M., Chabat, B., *Methodes de la Theorie des fonctions d'une variable complexe*, Mir, Moscou, 1977, (in French).

[74] Sveshnikov, A. G., Tikhonov, A. N., *Theory of Functions of a Complex Variable*, Mir, Moscow, 1979.

[75] Sidorov, Yu. V., Fedoryuk, M. V., Shabunin, M. I., *Lectures on the Theory of Functions of a Complex Variable*, Mir, Moscow, 1985.

[76] Fuchs, B. A., Shabat, B. V., *Functions of a Complex Variable and some their Applications*, Pergamon Press, Oxford, 1964.

[77] Lunts, G. L., Eltsgolts, L. E., *Functions of a Complex Variable*, Fizmatgiz, Moscow, 1958, (in Russian).

[78] Bitsadze, A. V., *Basics of the Theory of Analytic Functions of a Complex Variable*, Nauka, Moscow, 1984, (in Russian).

79 Rothe, E. R., Ollendorff, F., Pohlhausen, K., *Theory of Functions Applied to Engineering Problems*, Massachusetts Institute of Technology, Cambridge (Mass), 1933.

80 Phillips, E. G., *Functions of a Complex Variable with Applications*, Olyver and Boyd, Edinburgh-London, 1946.

81 Green, S. L., *The Theory and Use of the Complex Variable*, Isaac Pitman, London, 1953.

82 Churchill, R. V., *Complex Variables and Applications*, McGraw-Hill, New York, 1960.

83 Nehari, Z., *Introduction to Complex Analysis*, Allyn and Bacon, Boston, 1968.

84 Walker, M., *Conjugate Functions for Engineers*, Dover, New York, 1964.

85 Cunningham, J., *Complex-Variable Methods in Science and Technology*, Van Nostrand, New York, 1965.

86 Carrier, G. F., Krook, M., Pearson, C. E., *Functions of a Complex Variable, Theory and Technique*, McGraw-Hill, New York, 1966.

87 Henrici, P., *Applied and Computational Complex Analysis*, V. 1–3, Wiley, New York, 1974.

88 Fricke, R., *Elliptische Funktionen und ihre Anwendungen*, Bd. 1–2, Teubner, Berlin-Leipzig, 1916, 1922.

89 Fricke, R., Elliptische Funktionen, in *Enzyklopaedie der mathematischen Wissenschaften*, Bd. 2, Teil 2, Teubner, Leipzig, 1913, S. 177–348.

90 Burkhardt, H., Elliptische Funktionen, in *Funktionentheoretische Vorlesungen*, Bd. 2, Gruyter, Berlin, 1920.

91 Koenig, R., Krafft, M., *Elliptische Funktionen*, Gruyter, Berlin, 1928.

92 Tricomi, F., *Elliptische Funktionen*, Akademische Verlags, Leipzig, 1948.

93 Oberhettinger, F., Magnus, W., *Anwendungen der elliptischen Funktionen in Physik und Technik*, Springer, Berlin, 1949.

94 Graeser, E., *Einfuehrung in die Theorie der elliptischen Funktionen und deren Anwendungen*, Oldenbourg, Munuch, 1950.

95 Milne-Thomson, L. M., *Jacobian Elliptic Functions Tables*, Dover, New York, 1950.

96 Akhiezer, N. I., *Elements of the Theory of Elliptic Functions*, Nauka, Moscow, 1970, (in Russian).

97 Zhuravskii, A. M., *Handbook on Elliptic Functions*, Acad. of Sci. Publ., Moscow, 1941, (in Russian).

[98] Byrd, P. F., Friedman, M. D., *Handbook of Elliptic Integrals for Engineers and Physicists*, Springer, Berlin, 1954.

[99] Bateman, H., Erdelyi, A., *Higher Transcendental Functions*, McGraw-Hill, New York, 1955.

[100] Golubev, V. V., *Lectures on Analytical Theory of Differential Equations*, Gostekhizdat, Moscow-Leningrad, 1950, (in Russian).

[101] Tikhonov, A. N., Samarskii, A. A., *Methods of Mathematical Physics*, Nauka, Moscow, 1977, (in Russian).

[102] Lavrent'ev, M. A., *Conformal Mappings with Applications to Some Problems of Mechanics*, Gostekhizdat, Moscow, 1946, (in Russian).

[103] Solomentsev, E. D., *Functions of a Complex Variable and Their Applications*, Vysshaja Shkola, Moscow, 1988, (in Russian).

[104] Lavrik, V. I., Filchakova, V. P., Jashin, A. A., *Conformal Mappings of Physics-Topological Models*, Naukova Dumka, Kiev, 1990, (in Russian).

[105] Holzmueller, G., *Einfuerung in die Theorie der isogonalen Verwandschaften und der konformen Abbildungen mit Anwendungen auf mathematische Physik*, Teubner, Leipzig, 1882.

[106] Bateman, H., *Partial Differential Equations of Mathematical Physics*, Cambridge University Press, 1932.

[107] Morse, P. M., Feshbach, H., *Methods of Theoretical Physics*, V. 1–2, McGraw-Hill, New York, 1953.

[108] Betz, A., *Konforme Abbildung*. Springer, Berlin, 1964.

[109] Kochin, N. E., Kibel', I. A., Roze, N. V., *Theoretical Hydromechanics*, V. 1, Fizmatgiz, Moscow, 1963, (in Russian) or Kotschin, N. J., Kibel, I. A., Rose, N. W., *Theoretische Hydromechanik*, Bd. 1–2, Acad. Verlags, Berlin, 1954, (in German).

[110] Sedov, L. I., *Two-dimensional Problems in Hydrodynamics and Aerodynamics*, Wiley, New York, 1965.

[111] Lavrentiev, M. A., Shabat, B. V. *Problems of Hydrodynamics and Their Mathematical Models*, Nauka, Moscow, 1977, (in Russian).

[112] Lamb, N., *Hydrodynamics*, Cambridge Univ. Press, 1932.

[113] Miln-Thomson, L. M., *Theoretical Hydrodynamics*, Macmillan, London — New York, 1960.

[114] Golubev, V. V., *Theory of Airplane Wing in Planeparallel Flow*, Gostekhizdat, Moscow, 1938, (in Russian).

[115] Golubev, V. V., *Lectures on the Aerofoil Theory*, Gostekhizdat, Moscow, 1950, (in Russian).

[116] Carafoli, E., *Aerodynamique des Ailes d'Avion*. Etienne Chiron, Paris, 1928.

[117] Carafoli, E., *Tragefluegeltheorie*, Verlag Technik, Berlin, 1954.

[118] Glauert, H., *The Elements of Aerofoil and Airscrew Theory*, Cambridge Univ. Press, London, 1948.

[119] Milne-Thomson, L. M., *Theoretical Aerodynamics*, MacMillan, London, 1948.

[120] Kochin, N. E., *Hydrodynamical Theory of Grids*, Gostekhizdat, Moscow, 1949, (in Russian).

[121] Deich, M. E., Samoilovich, G. S., *Basics of Aerodynamics of Shaft Turbo-Machines*, Mashgiz, Moscow, 1959, (in Russian).

[122] Zhukovskii, M. I., *Calculation of Streaming Around Grids of Profiles of Turbo-Machines*. Mashgiz, Moscow, 1960, (in Russian).

[123] Stepanov, G. Yu., *Hydrodynamics of Turbo-Machine Grids*, Fizmatgiz, Moscow, 1962, (in Russian).

[124] Fil'chakova, V. P., *Conformal Mappings of Domains of Special Type*, Naukova Dumka, Kiev, 1972, (in Russian).

[125] Weinig, F., *Die Stroemung um Schaufeln von Turbomaschinen*, Barth, Leipzig, 1935.

[126] Gostelow, J. P., *Cascade aerodynamics*, Pergamon Press, New York, 1984.

[127] Nelson-Skornjakov, F. B., *Filtration in homogeneous medium*, Sovetskaja Nauka, Moscow, 1949, (in Russian).

[128] Polubarinova-Kochina, P. Ja., *Theory of Ground Water Motion*, Nauka, Moscow, 1977, (in Russian).

[129] Fil'chakov, P. F., *Theory of Filtration under Hydrotechnical Constructions*, V. 1–2, Acad of Sci. Publ., Kiev, 1959–1960, (in Russian).

[130] Birkhoff, G., Zarantonello, E. H., *Jets, Wakes and Cavities*, Academic Press, New York, 1957.

[131] Gurevich, M. I., *Theory of Jets of the Ideal Liquid*, Nauka, Moscow, 1979, (in Russian).

[132] Asnin, I. M., *Calculations of Electromagnetic Fields (Planeparallel Field)*, Leningrad, 1939, (in Russian).

[133] Mirolubov, N. N., Kostenko, M. V., Levinshtein, M. V., Tikhodeev, N. N., *Methods of Calculations of Electrostatic Fields*, Vysshaja Shkola, Moscow, 1963, (in Russian).

[134] Govorkov, V. A., *Electrical and Magnetic Fields*, Energija, Moscow, 1968, (in Russian).

[135] Kovalev, I. S., *Theory and Calculation of Strip-Line Waveguides*, Nauka i Tekhnika, Minsk, 1967, (in Russian).

[136] *Strip-Lines and UHF Systems*, Sedykh, V. M., Ed., Vysshaja Shkola, Khar'kov, 1974, (in Russian).

[137] Javor, S. Ja., *Focusing of Charged Particles by Quadrupole Lenses*, Atomizdat, Moscow, 1968, (in Russian).

[138] Ollendorff, F., *Potentialfelder der Electrotechnik*, Springer, Berlin, 1932.

[139] Buchholz, H., *Elektrische und magnetische Potentialfelder*, Springer, Berlin, 1957.

[140] Smythe, W. R., *Static and Dynamic Electricity*, McGraw-Hill, New York, 1950.

[141] Gibbs, W. J., *Conformal Transformations in Electrical Engineering*, Chapman and Hall, London, 1958.

[142] Bewley, L. V., *Two-Dimensional Fields in Electrical Engineering*, Dover, New York, 1963.

[143] Binns, K. J., Lawrenson, P. J., *Analysis and Computation of Electric and Magnetic Field Problems*. Pergamon Press, Oxford, 1963.

[144] Weber, E., *Electromagnetic Theory, Static Field and Their Mapping*, Dover, New York, 1965.

[145] Collin, R. E., *Foundations for Microwave Engineering*, McGraw-Hill, New York, 1966.

[146] Hawkes, P. W., *Quadrupoles in Electron Lens Design*, Acad. Press, London, 1970.

[147] Kolosov, G. V., *Applications of Complex Diagrams and Theory of Functions of a Complex Variable to the Theory of Elasticity*, Gostekhizdat, Moscow, 1935, (in Russian).

[148] Muskhelishvili, N. I., *Some Basic Problems of the Mathematical Theory of Elasticity*, P. Noordhoff, Groningen, 1953.

[149] Savin, G. N., *Concentration of Stresses near Holes*, Gostekhizdat, Moscow, 1951, (in Russian).

[150] Novozhilov, V. V., *Theory of Elasticity*, Sudpromgiz, Moscow, 1958, (in Russian).

[151] Kuliev, S. A., *Two-Dimensional Problems of Elasticity Theory*, Stroiizdat, Moscow, 1991, (in Russian).

[152] Timoshenko, S. P., Goodier, J. N., *Theory of Elasticity*, McGraw-Hill, New York, 1970.

[153] Sokolnikoff, I. S., *Mathematical Theory of Elasticity*, McGraw-Hill, New York, 1956.

[154] Wang, C.-T., *Applied Elasticity*, McGraw-Hill, New York, 1953.

[155] Green, A. E., Zerna, W., *Theoretical Elasticity*, Clarendon Press, Oxford, 1968.

[156] Sneddon, I. N., Berry, D. S., *The Classical Theory of Elasticity*, Springer, Berlin, 1958.

[157] Weber, C., Guenther, W., *Torsionstheorie*, Acad.-Verlag, Berlin, 1958.

[158] Miln-Thomson, L. M., *Plane Elastic Systems*, Springer, Berlin, 1960.

[159] Miln-Thomson, L. M., *Antiplane Elastic Systems*, Springer, Berlin, 1962.

[160] England, A. H., *Complex Variable Methods in Elasticity*. Wiley, London, 1971.

[161] Dugdale, D. S., Ruiz, C., *Elasticity for Engineers*, McGraw-Hill, London, 1971.

[162] Hahn, H. G., *Elastizitaetstheorie*, Teubner, Stuttgart, 1985.

[163] Carslaw, H. S., Yaeger, J. C., *Conduction of Heat in Solids*, Clarendon Press, Oxford, 1959.

[164] Volkovyskii, L. I., Lunts, G. L., Aramanovich, I. G., *A Collection of Problems on Complex Analysis*, Pergamon Press, New York, 1965.

[165] *A Collection of Problems on the Theory of Analytic Functions*, Evgrafov, M. A., Ed., Nauka, Moscow, 1972, (in Russian) or *Recuell des problemes sur la theorie des fonctions analytiques*, Evgrafov, M. A., Ed., Editions Mir, Moscow, 1974, (in French).

[166] Lebedev, N. N., Skal'skaja, I. P., Ufljand, Ja. S., *A Collection of Problems on the Mathematical Physics*, Gostekhizdat, Moscow, 1955, (in Russian).

[167] Leont'eva, T. A., Panferov, V. S., Serov, V. S., *Problems of the Theory of Functions of a Complex Variable*, Izdat. MGU, Moscow, 1992, (in Russian).

[168] Krzyz, J., *Problems in Complex Variables Theory*, Amer. Elsevier Publ. Co., New York, 1971.

[169] Spiegel, M. R., *Theory and Problems of Complex Variables*, McGraw-Hill, New York, 1964.

[170] Lavrik, V. I., Savenkov, V. N., *Handbook on Conformal Mappings*, Naukova Dumka, Kiev, 1970, (in Russian).

[171] Kober, H., *Dictionary of Conformal Representations*, Dover, New York, 1952.

[172] Koppenfels, W., Stallmann, F., *Praxis der konformen Abbildung*, Springer, Berlin, 1959.

173 Jahnke, E., Emde, F., Loesch, F., *Tafeln hoeherer Funktionen*, Teubner, Stuttgart, 1960.

174 Korn, G. A., Korn, T. M., *Mathematical Handbook for Scientists and Engineers*, McGraw-Hill, New York, 1961.

175 Abramovitz, M., Stegun, I., *Handbook of Mathematical Functions*, National Bureau of Standards, New York, 1964.

176 Savelov, A. A., *Plane Curves*, Fizmatgiz, Moscow, 1960, (in Russian).

177 *Mathematical encyclopedia*, V 1–5, Sovetskaja encyclopedia, Moscow, 1977–1985, (in Russian).

Index